UNDER THE WEATHER:
COVID-19 BIOSOCIAL
SYSTEM DYNAMICS

UNDER THE WEATHER: COVID-19 BIOSOCIAL SYSTEM DYNAMICS

AUSTIN MARDON, TINA WU,
GRACE LETHBRIDGE, LILY LIU, MIRAY MAHER,
& CATHERINE MARDON

GM PRESS

TABLE OF CONTENTS

CHAPTER 1
INTRODUCTION

BY TINA WU

Viruses have existed for billions of years, far surpassing the dawn of human existence. The estimated number of viruses on Earth is staggeringly large at more than a quadrillion quadrillion—there are more viruses on Earth than stars in the entire universe (Wu). It comes by no surprise that encounters with disease is a large aspect of human thought and experience. Viruses touch on all aspects of human life; it is represented in culture, politics, economics, and society. The popular children's song "Ring around the Rosy" has a hidden dark undertone of pestilence and death. In Ontario, almost every worker has the benefit of three sick days per calendar year ("Sick Leave"). Almost all apocalyptic media, including zombie movies, feature a virus. Viruses are a social phenomena as much as they are a physiological one, and manifest in all areas of our lives.

Another focal outlook on viruses is the biological one. Various scientific lenses can be taken to examine the complexities of viral disease: evolutionary history, environment and geography, statistics and epidemiology, bioinformatics, symptomatology, and of course, virology. Compared to the Middle Ages when plague ravaged the population and there was little hope for the seriously ill, modern medicine has drastically diminished the threat of viruses. The development of vaccinations and treatments has allowed us to eradicate smallpox, rinderpest, and restrict polio by decreasing cases by 99% (Corona; "Does Polio Still Exist? Is It Curable?"). However, recent events have shown the dangers of viral infection are far from over. The intersection of the biological and social aspects of viruses are brought to light in the appearance of a new viral threat to the human population, with nearly 20 million cases worldwide as of early August 2020 ("Canada").

In December of 2019, a novel coronavirus surfaced in the wet markets of Wuhan, China. The virus affected the lower and upper respiratory tracts of humans, with potential impacts on various other organ systems including the circulatory and renal systems. Symptoms of the virus are difficult to distinguish from those of the flu; fever, dry cough, and fatigue are most common ("Q&A on coronaviruses (COVID-19)"). Since its initial outbreak, a series of events have led the coronavirus to become the largest pandemic of the 21st century.

The identification of this novel virus was delayed by political factors. BBC reports that in late December, Dr. Li Wenliang noticed that seven of his patients had cases of a virus that looked like severe acute respiratory syndrome (SARS), which caused a pandemic in 2003. He alerted his fellow medics about the outbreak on December 30 and advised them to wear protective clothing to avoid infection. Four days later, he was visited by authorities from the Public Security Bureau and signed a letter to stop "making false comments". It was January 20, 2020, when the Chinese government finally declared the outbreak an emergency. Unfortunately, Dr. Li Wenliang contracted the virus himself and passed away on February 7 (Hegarty). He is praised as a hero for raising the alarm on the new outbreak with his death sparking outrage worldwide and drawing government and healthcare systems' attention to the novel coronavirus internationally. The events that followed include the imposition of strict lockdowns in China, infection of a cruise ship in Japan, and a number of outbreaks in Iran, Italy, and other countries (Schumaker). The virus gained traction and on March 11, 2020, coronavirus disease 2019 (COVID-19) was officially declared a pandemic by the World Health Organization (WHO).

An infectious disease that is limited to a community, region, or country is known as an epidemic. When a disease such as COVID-19 changes from an epidemic to a pandemic, this does not have to do with the severity of the disease: rather, the term reflects its geographical spread. A pandemic occurs when an infectious disease spreads worldwide, crossing international borders, and usually afflicting a large number of people ("The Classical Definition of a Pandemic Is Not Elusive"). The declaration of a pandemic is significant even though the label itself holds no legal implications because of the fear and power we associate with the word. The WHO was greatly scrutinized for declaring the swine flu (H1N1 virus) a pandemic in 2009 because it led to widespread panic as governments spent a lot of money on vaccine research for a disease that ended up being fairly mild and easy to contain. Unnecessary panic has many downsides, but declaring a pandemic can also allow the general public to mentally and physically prepare for the spread of the disease, potentially aiding in mitigation of viral spread and economic losses (Brown and Moffit).

The Biological Side of COVID-19

The COVID-19 pandemic is caused by a virus named severe acute respiratory syndrome coronavirus 2 (SARS-CoV-2), which belongs to the family of coronaviruses. Many different strains of coronaviruses circulate among a diverse range of animal species like cows, bats, birds, dogs, cats, and pigs (Shi et al.; Saif). These viruses cause symptoms varying from respiratory to gastrointestinal. Understanding the evolutionary history of coronaviruses in animal populations can help us discern how mutations that cause zoonotic transfer, allowing the virus to jump from species to species, occur. With SARS-CoV-2, researchers speculate that zoonotic transfer occurred from bats or perhaps spread from bats to an intermediate host which was then contracted by humans (Fox). Additionally, a firm grasp on the mechanisms of infection and symptomology in animals such as livestock and pets that are in contact with humans may help us better predict their biological impact on humans.

Coronaviruses have crossed into the human population seven times in recent years. They all share a similar structure, with spike-like glycoproteins that aid in the invasion of their host's cells. However, some cause the typical symptoms accompanying an upper respiratory infection (i.e the common cold), while others are deadlier. Four of the coronaviruses that infect humans cause the common cold. Two other viruses have caused deadly epidemics in the 21st century. These are SARS and the Middle East respiratory syndrome (MERS), which had outbreaks in 2002 and 2012 respectively. SARS-CoV-2 now joins this list ("Coronavirus"). MERS-CoV, SARS-CoV and now SARS-CoV-2 have caused significant fatalities and understanding the mechanisms by which they invade a host's body through symptoms and post-mortems will allow effective treatments to be implemented in the future.

As a highly infectious disease rapidly expanding across the globe, COVID-19 has shaken the world with its transmissibility. This sparked research on the optimum conditions for the transmission of the novel coronavirus, which involve both environmental and viral characteristics. Factors like air quality, humidity, temperature and land transformations all play a critical role in the virus's survival and transmission (Ma, Yueling, et al). Additionally, studying the molecular mechanisms of SARS-CoV-2 may give insight into the infectivity of the virus, especially compared to other coronaviruses such as SARS-CoV. The viral characteristics that enable SARS-CoV-2 to spread effectively are a highlight of current medical research.

The Sociological Side of COVID-19

While the highly contagious disease sparks a lot of fear, there is also hope and action. Luckily, the collective work of government systems, healthcare professionals, and the general population can help combat the spread of the disease. Accurate data collection and time-space models can help inform policymakers of the best containment strategies, whether it be social distancing, quarantining, contact tracing, border control, or education. These strategies are meant to buy time for researchers to find vaccinations and treatments while minimizing the death toll. Different countries will have different approaches to end the pandemic: delay and vaccinate is only one of the three. Racing through it, and coordinate and crush are two other tactics that are seen within the global community as we continue to press through the pandemic.

To combat COVID-19 in Ontario and Quebec, the provincial governments made the decision to shut down non-essential businesses in late March of 2020. Needless to say, this has drastic implications for the economy. Despite government relief for rent on commercial buildings, many businesses went bankrupt. Aid was also provided for individuals who lost their jobs due to COVID-19, known as the Canadian Emergency Response Benefit, while university students could apply for the Canadian Emergency Student Benefit. As the country plummets into recession, public spending rises dramatically. COVID-19 hit the travel and tourism industry the hardest due to restricted air travel and travel bans to thwart the spread of the virus. Furthermore, this pandemic has changed consumer patterns, allowing for the rise of eCommerce, contactless payment methods, and the decline in sales of many retail establishments. Lastly, the housing market faced a small decline leaving potential home buyers and sellers at a point of uncertainty. Economic recovery in Canada involves a controlled reopening of businesses to resume economic activity without initiating a second wave of COVID-19 cases.

As far as the economy is impacted, so are our social behaviours. To encourage physical distancing, elementary, secondary schools, and universities alike have closed and shifted to online learning. Gatherings of more than 5 people were prohibited in Ontario in late March, although restrictions have eased to 10 people in early June. Individuals are encouraged to stay at home when possible. In place of in-person interactions, viewing faces through a screen has become the norm. Social media usage in particular has gone up in recent times. According to one site, messaging across Facebook, Instagram and WhatsApp has increased 50% in countries hit the hardest by the virus as of late April of 2020. While social media is a fantastic tool to maintain communication with friends and family during a time that face-to-face meetings are less attainable. It can also be a good platform to educate the population on COVID-19. Unfortunately, issues arise when inaccurate information is posted. With social media, these non-fact-checked messages can spread like wildfire and reach a significant number of people. Fear-mongering is also prevalent in the media. It falls on the consumers to remain vigilant in ensuring the information they receive when they use social media platforms is reliable, researched, and factual.

Conclusion

COVID-19 is not a strictly biological or social phenomenon. The disease manifests itself at the centre of both. As crucial as the environmental, evolutionary, and physiological factors are, there are equally important social, political, and economic aspects to consider. Exploring the novel coronavirus through an interdisciplinary lens enables us to see the big picture of the pandemic to determine the steps that need to be taken to prevent a tragedy of this scale from befalling humanity again—but if it does, to ensure that we are prepared to take the threat head-on.

CHAPTER 2

THE RELATIONSHIP OF COVID-19 WITH OTHER VIRUSES

BY LILY LIU

Introduction

The coronavirus disease of 2019, more commonly abbreviated as COVID-19, began its outbreak in the epicentre of Wuhan, China and has since evolved into a mass-scale pandemic affecting the globe. Officially named by the World Health Organization on February 11, 2020, the disease has infected more than 13.5 million people and caused over half a million deaths worldwide as of July 2020.

As an introduction to the disease, the virus that causes COVID-19 is believed to spread through contact with respiratory droplets, which are primarily produced through coughing or sneezing. When a person who has been infected secretes these droplets, the people around them are exposed to the risk of the virus by breathing them in. Both symptomatic and asymptomatic hosts are able to transmit the virus, and its symptoms may range anywhere from mild discomfort in the chest to severe respiratory infections, which may eventually cause serious illness or death.

COVID-19 is not the first of its kind. Coronaviruses are a common family of viruses - named for their unique crown-shaped structure, or "corona" - that are mainly known for causing mild respiratory illnesses such as the common cold. They were first discovered in domestic animals during the 20th century, and have the ability to spread between organisms by sticking to surfaces and subsequently entering the system to infect the host. In this particular case, the strain of virus causing COVID-19 was a zoonotic pathogen originally solely transmitted between animal populations, and only became the international crisis it is today when it began to be contagious among humans, resulting in a rapid spread that has gone unchecked due to the lack of a vaccine or natural immunity.

Although COVID-19 has been classified as a "novel" coronavirus - meaning a coronavirus that has not been previously identified, therefore possessing unique and unknown characteristics and properties - the disease shares a number of similarities and relations to other viruses that have been affecting the human population long before COVID-19's discovery. These include the common cold, the SARS epidemic, the Middle East respiratory syndrome or MERS, all of which can be said to display significant links to the virus behind the current COVID-19 pandemic. On the basis of biology, viruses can be said to develop and evolve through their interactions with the organisms they are able to affect. Because of this, learning about COVID-19 and its relationships with other previously discovered viruses are crucial; understanding how similar diseases form and spread may provide a wealth of information regarding the nature of the virus in question, allow us to create connections with the cases of the past, and lead to a deeper comprehension of the potential ways that COVID-19 may be controlled and contained.

Out of the numerous strains of coronavirus that have been discovered in infected organisms, only seven are known to affect humans. Of these seven, four types - specifically 229E, NL63, OC43, and HKU1 - are classified as "common" human coronaviruses that cause milder symptoms such as the common cold. The other three, however, are able to cause much more severe illnesses that may become fatal. These include the MERS (Middle East Respiratory Syndrome) coronavirus, SARS-CoV, and SARS-CoV-2, which causes the disease known as COVID-19 today.

Relationships with COVID-19

Common Cold

The most common type of human coronaviruses cause effects that are similar to COVID-19 in the form of the common cold. Although these diseases possess significant differences both in terms of structure and intensity, they share a certain amount of similarities in regards to their symptoms - the common cold, like COVID-19, can be diagnosed through sneezing, coughing, breathing difficulties, a sore throat, and in rare cases, a fever. The key distinction between these common coronaviruses and COVID-19 is their degree of severity. While patients with a common cold usually only suffer from mild respiratory infections and tend to recover after a rather short period of time, patients who have been infected with the coronavirus causing COVID-19 have been recorded to experience far graver symptoms, some of which may result in serious illness or death.

COVID-19, at first glance, is nearly indistinguishable from the common cold. Regardless of its more severe cases, being infected with the virus that has caused the international pandemic today results in mild to moderate symptoms in a majority of situations. This is due to the fact that, at its core, the virus behind COVID-19 was born from a mutation of one of the common coronaviruses. This mutation allowed it to develop into a more dangerous and deadly version of the original virus.

Although it is still being extensively studied by scientists around the world, it is generally believed that the disease originated when a coronavirus that was known to only infect animals evolved, allowing it to become able to infect and spread among humans - hence the term "novel" that is used to refer to the virus behind COVID-19, which distinguishes it as a previously undiscovered virus that only came into being in the year 2020. While much debate still surrounds the nature of the disease and its formation, it is generally believed that the origin of the virus can be traced back to live animal markets in Wuhan, China, the epicentre of the pandemic. The theory links the virus that affects humans to a virus that was initially only carried by bats. This would support the assumption that the virus behind COVID-19 was a zoonotic pathogen that eventually mutated to be able to transmit itself and infect the human population.

As a result of its direct connection to the more mild forms of coronavirus, the relationship between COVID-19 and the virus that causes the common cold is fairly straightforward.

MERS (Middle East Respiratory Syndrome)

The Middle East Respiratory Syndrome, more commonly abbreviated as MERS, is a respiratory disease caused by the MERS coronavirus. It was first discovered and reported in Saudi Arabia in 2012 and has since been recorded to spread to countries around the globe, although a majority of the cases are concentrated in Western Asia within the Arabian Peninsula. According to the World Health Organization, there have been nearly 2,500 confirmed cases of MERS since its first outbreak in September 2012.

Similar to COVID-19, MERS is a viral respiratory illness caused by a coronavirus that is capable of infecting humans. Symptoms include coughing, fevers, chills, and difficulty in breathing, all of which are shared with COVID-19. MERS is also largely transmitted through respiratory droplets, although it may also spread from aerosols and direct contact from person to person as a result of close proximity with an infected patient. Cases tend to be more severe in patients who have already been diagnosed with health conditions or chronic illnesses. The Center of Disease Control (CDC) reports that about 3 to 4 out of every 10 people diagnosed with MERS have died, giving it a far higher death rate than COVID-19, which has only been reported to cause 3 deaths out of 100 confirmed patients.

The coronavirus behind MERS - much like the one behind COVID-19 - is believed to have originated within species of dromedary camels, which are commonly seen in Saudi Arabia, the epicentre of the disease. This is a shared similarity with the virus that causes COVID-19, which has been theorized to have passed from an infected bat to a human through contact with the animal. These connections between MERS and COVID-19 further exemplify the idea that coronavirus infections capable of affecting humans only come to be as a result of a mutation from a typical coronavirus, which have been shown to solely affect animal populations.

SARS (Severe Acute Respiratory Syndrome)

The Severe Acute Respiratory Syndrome, also known as SARS, is another example of a more serious respiratory illness that affects patients on the basis of a viral infection. First discovered in Guangdong, China in the year 2002, SARS has been recorded in more than 8,000 patients around the world and has resulted in nearly 800 deaths. The epidemic lasted for a total of eight months before the World Health Organization officially announced that the global outbreak had been contained in July 2003.

The coronavirus that causes SARS has been classified as SARS-CoV. Symptoms include a high fever, headaches, overall feelings of discomfort and pain in the body, mild respiratory illness, and gastrointestinal disturbances such as diarrhea. Confirmed patients have also been recorded to develop pneumonia as a result of being infected. SARS is spread through close contact between people - like COVID-19, respiratory droplets as a result of sneezing or coughing are the main mediums for transmission, as well as contact with contaminated surfaces or objects. What is notable is that no known cases of SARS transmission were reported to have developed before the infected patient began showing symptoms, which is different from COVID-19's ability to be transmitted by asymptomatic patients as well as symptomatic ones. Both coronaviruses have been shown to have similar aerosol and surface stability when studied under a microscope. In comparison to COVID-19's 3-4% mortality rate, however, SARS has been shown to have a more severe mortality rate of about 10%.

SARS is believed to have mutated from a coronavirus carried by bats, which is identical to the theory behind the origin of COVID-19 - but an intermediate host in the form of civet cats was involved in the case of SARS, meaning that the bats infected the cats first, leading to the virus spreading to humans through the interactions of wet markets. These markets were also organized in China, the home of the epicentres of both the SARS epidemic and the COVID-19 pandemic. The response to the global SARS outbreak involved nearly identical tactics as the current response to the 2020 COVID-19 outbreak, including travel restrictions, mandatory quarantine and containment, and issuing city-specific alerts to stop transportation.

Influenza

Influenza, more known by its common name "the flu", is a type of disease known to cause seasonal epidemics that usually emerge in the later months of the year. The four main types of influenza viruses are categorized by the letters A, B, C, and D, wherein A and B are the typical human viruses that result in the yearly wave of illness found among the world population. While these yearly trends are common, influenza has also been recorded to cause mass international epidemics on the scale of the current COVID-19 pandemic, most notably during the "Spanish Flu" pandemic in 1918, the "Asian Influenza" in 1957, and more recently, the 2009 H1N1 pandemic that has also been referred to as the "Swine Flu". To date, influenza outbreaks have been shown to cause an average of 3 - 5 million severe cases and more than 500,000 deaths per year.

Similar to the other viruses that have been previously discussed in this chapter, influenza affects the human respiratory system and causes symptoms that resemble those of the common cold, including fevers, dry coughs, sore throats, aching muscles, a runny nose, and bodily fatigue. The virus may also be spread through the inhalation of respiratory droplets produced by sneezing or coughing, as well as physical contact with the droplets through the eyes, nose, or mouth. Furthermore, the influenza virus can be found and transmitted to humans through a wide variety of animal hosts, including chickens, pigs, ducks, horses, and aquatic mammals such as whales and seals. This bears resemblance to the nature of COVID-19, which has been theorized to have originated as a virus that infects bats before its eventual mutation.

Perhaps the largest distinction between the influenza virus and the coronavirus causing COVID-19 is its ability to infect and spread without barriers. Unlike the new case of the COVID-19 pandemic, influenza is a recurring, seasonal epidemic that emerges year by year, which has allowed previously infected hosts to build up a gradual residual immunity to the virus that prevents them from being re-infected. Additionally, despite the slight mutations and changes in the new strains of the influenza virus that appear each year, there are effective vaccines and anti-flu treatments in place to support new patients and reduce the severity of their illness.

CHAPTER 3

SYMPTOMS OF CORONAVIRUS INFECTIONS IN HUMANS

BY MIRAY MAHER

Introduction

COVID-19 belongs to a viral family called Coronaviridae. These viruses are zoonotic; they are transmitted between animals and humans. For example, MERS was determined to have been spread to humans through dromedary camels, SARS through civet cats and the novel coronavirus (COVID-19) through bats (WHO Africa). These coronaviruses share similar characteristics like having distinct spike-like glycoproteins that resemble a crown, hence the name coronavirus (Auwaerter). The greatest variety of coronaviruses are present in bats, making them the most likely natural reservoir for these viruses, with peridomestic animals (animals living around humans) acting as a bridge that may transfer these zoonotic coronaviruses from wildlife to humans (Paules et al.). However, although they share similar structural characteristics, different strains target different parts of the body presenting as various symptoms and symptoms differ from species to species. This chapter will explore the effects that different types of coronaviruses have on the human body and how some eventually cause the demise of their victims.

Human Coronaviruses:

There are seven types of coronaviruses that infect humans. Four of them are endemic globally (regularly found among people) and account for 10%-30% of upper respiratory tract infections in adults (Paules et al.), making them the second leading cause for the common cold after rhinoviruses (Poutanen). These four HCoVs (human coronaviruses) are 229E, NL63, OC43, HKU1. These viruses produce very mild symptoms in most patients therefore have received minimal attention. The other three are more serious and even fatal.

MERS-CoV, or the virus that causes Middle East Respiratory Syndrome (MERS), made its first appearance in Saudi Arabia in 2012. The natural reservoir for this coronavirus is bats but camels are believed to have been the intermediary link that transferred this disease over to humans (Paules et al.). Having originated there, most cases are localized around Saudi Arabia with a global case count of 2494 and 858 deaths. Out of the three most fatal coronaviruses, MERS has by far the highest mortality rate of 34.4% (Petrosillo et al.).

23

Severe Acute Respiratory Syndrome, caused by the coronavirus SARS-CoV first appeared in the Guangdong province of Southern China in 2002. By the end of 2003, there were over 8000 cases and 774 deaths worldwide in 26 countries (Peiris et al.). In 2004, a "SARS-like virus" was isolated from civet cats in China, identifying the origin of the outbreak (CDC Contributors). The mortality rate for SARS is 10%, significantly lower than the more recent MERS (Paules et al.).

Lastly, SARS-CoV-2, which causes COVID-19, is the virus responsible for the current pandemic. As of early August, 2020, this pandemic has over 15.5 million cases worldwide to its name and about six hundred thousand deaths (Worldometer.com). The first reported case was in December 2019 in Wuhan (Paules et al.) and is linked to a viral spillover event from bats to some intermediate host in a wet market that transmitted the disease to humans. Although it seems that COVID-19 is not as fatal as MERS and SARS, the totality of this pandemic's implications is yet to be anticipated (Paules et al.).

All human coronaviruses are typically transmitted through respiratory droplets that are ejected through the nose and mouth (WHO Africa). The next section will go into depth about the symptoms of each of these viruses, their similarities and differences.

229E, NL63, OC43, HKU1 Symptoms:

These endemic coronaviruses often clinically present with the typical common cold symptoms of rhinorrhea (runny nose), sneezing, coughing, sore throat, fatigue, nasal congestion and a low-grade fever (Poutanen). Occasionally, they may result in more serious complications like pneumonia and bronchiolitis (inflammation of bronchioles) in higher-risk patients and some, in particular, can cause even gastrointestinal and neurological issues (Liu et al.).

229E

HCoV-229E was first isolated in 1966 and is believed to have originated from African bats and have been passed to humans through camels.

24

In addition, HCoV-229E shares many evolutionary characteristics with the much more recent and deadly virus, MERS-CoV (Liu et al.)

This virus's incubation period is about 2-5 days, with illness lasting afterwards 2-18 days. This virus is more prevalent during the winters of temperate climates areas (Liu et al.).

In healthy adults, HCoV-229E infection is associated with common cold symptoms such as headache, nasal discharge, general fatigue and a sore throat and in some patients, a fever may be present. However, infants, children, seniors, and immunocompromised patients are more likely to suffer from severe lower respiratory tract infections, including bronchiolitis and pneumonia. Furthermore, HCoV-229E can cause acute otitis media (inflammation of the middle ear) and exacerbations in asthma (Poutanen). There is also a small link between HCoV-229E and the development of Kawasaki disease (Liu et al.). Kawasaki disease is an acquired disease that afflicts children five years old or younger and it causes inflammation of the arteries, capillaries and veins (Cafasso).

NL63

HCoV-NL63 was first isolated from a seven-month-old girl in the Netherlands in 2004. She presented with coryza (inflammation of the nasal mucous membranes), conjunctivitis (pink eye), fever and bronchiolitis (Liu et al.).

In most patients, HCoV-NL63 infection is associated with mild cold-like symptoms such as fever, cough, sore throat and rhinitis (Liu et al.). However, HCoV-NL63 and HCoV-OC43 cause most of the respiratory infections leading to hospitalization and HCoV-NL63 causes a higher occurrence of lower respiratory infection (ex. bronchiolitis) than the other endemic coronaviruses (Abdul Rasool and Fielding). Often, HCoV-NL63 infects patients in conjunction with other viruses. 71% of patients infected with HCoV-NL63 also present with other respiratory viruses such as human rhinovirus, parainfluenza virus and enterovirus (Liu et al.). A weak link was also found between this virus and Kawasaki disease (Abdul Rasool and Fielding). Moreover, HCoV-

NL63 is one of the main causes of croup in children (Poutanen). Croup is a viral infection that affects children mostly under five. It causes swelling around the larynx (box-like structure holding vocal cords) that causes a cough that sounds similar to a barking seal (Selner). HCoV-NL63 has also been found to exacerbate asthma and cause febrile seizures (Chiu et al.).

Although HCoV-NL63 infects people of all ages, children under the age of five constitute the largest demographic affected by the virus. Around 1-10% of all cold-like infections are caused by this virus specifically (Liu et al.). It is more prevalent during the winter seasons within areas of a temperate climate and its incidence is globally distributed (Abdul Rasool and Fielding).

HKU1

HCoV-HKU1 was first identified in Hong Kong in 2005 from an adult with a chronic pulmonary disease (Liu et al.). At examination, the patient presented with rhinorrhea, cough with sputum, nasal congestion, fever, sore throat, and enlarged tonsils. HCoV-HKU1 infections have a much higher incidence of febrile seizures in children than other respiratory virus infections (Lau et al.). Moreover, HCoV-HKU1 has been also specifically associated with symptoms of gastroenteritis like vomiting and diarrhea (Poutanen).

HCoV-HKU1 affects people all over the globe, just like the other three mild HCoVs and is relatively frequent in adults during the spring and summer period (usually after the influenza season) (Liu et al.).

OC43

This virus was first isolated in 1967. As compared to HCoV-229E infections, sore throat manifestations occur more often during HCoV-OC43 infections, however, coryza (inflammation of nasal mucous membranes) occurs more often during the latter (Liu et al.).

This virus shares many of the same clinical manifestations as the other three mild coronaviruses, like common cold-like symptoms, however, HCoV-

OC43 has also been shown to have neuroinvasive properties. Experiments on mice have shown that this virus can infect neurons and cause encephalitis. In addition, an HCoV-OC43 epidemic that occurred in Normandy, France during 2001 found that 57% of those infected also experienced a wide range of gastrointestinal problems like diarrhea, abdominal pain and vomiting (Cuffari).

HCoV-OC43 is primarily prevalent during the winter months of areas of temperate climates (Liu et al.)

SARS

SARS-CoV's common symptoms include fever, cough, dyspnea and sometimes even diarrhea and the incubation period of SARS is 2 to 7 days. . 20% to 30% of cases require mechanical ventilation. Unfortunately, complications due to mechanical ventilation commonly occurred in the form of pneumothoraces (Poutanen). As previously mentioned, SARS has a 10% fatality rate with most deaths occurring in older patients with previous medical conditions. For patients over 60 infected with SARS, the mortality rate goes up to 50% (Poutanen).

The reason this virus is more deadly than the four endemic cold-causing viruses is because of the higher incidence of severe lower respiratory infections such as pneumonia. This occurs because the predominant human receptor for the SARS glycoproteins is the human angiotensin-converting enzyme 2 (ACE2) which is found primarily in the lower respiratory system instead of within the upper airway (Paules et al.). In addition, patients often presented with lower white blood cell count than normal, a condition called lymphopenia (Poutanen).

Infants and children under 12 are less afflicted with this virus, however, those who are infected develop much milder symptoms than their older counterparts. These symptoms are those of a typical cold (fever, cough, and rhinorrhea). Lymphopenia is usually also less severe and symptoms resolve more quicker in general. Adolescents and teenagers who got infected with SARS had more severe symptoms, similar to the adults. They developed fever, muscle aches, chills, dyspnea, and hypoxemia (Poutanen).

Lastly, pregnant women who caught SARS-CoV were at a higher risk for preterm delivery, miscarriage and the babies had a higher chance of gastrointestinal complications soon after birth such as necrotizing enterocolitis, however, there was no evidence that the SARS infection was transmitted to them from the mother (Poutanen).

Death by SARS is caused by the virus attacking the epithelial cells of the alveoli. Post-mortem examinations of SARS victims show that those who died after two weeks of the onset of infection did not have any SARS-CoV in lung tissue. However, those who have died before two weeks of infection were found to have SARS-CoV in pneumocytes which are the cells that line the alveoli. In all fatalities, the SARS virus was not found in any organ tissue except the lungs. This provides some pathological clues on how death by SARS occurs. Firstly, it seems that the immune system is capable of stopping viral proliferation after two weeks of infection however, by then, the extent of lung damage is so great that death is imminent even without continued viral replication. Therefore, death occurs only due to the SARS-CoV virus replicating within the lungs. (Nicholls et al.).

MERS

MERS causes many similar conditions as SARS such as severe pneumonia, however, MERS has distinct differences that make it in most cases, even deadlier. One of these factors is a longer incubation period ranging from 2 to 14 days with an average of 5 days.

MERS-CoV causes symptoms similar to that seen in SARS-CoV such as fever, myalgia, non-productive cough, and dyspnea. However, unlike other viral infections, symptoms associated with upper respiratory diseases such as rhinorrhea and a sore throat are usually absent (Poutanen). One of the key features of MERS is the prominent gastrointestinal and renal symptoms that differentiate it from other coronaviruses. The MERS-CoV virus possesses a special glycoprotein that has a high affinity for a protein called dipeptidyl peptidase 4 (DPP4). This protein is present in the lower respiratory system,

GI tract and the kidneys. This causes the digestive symptoms of diarrhea, vomiting and abdominal pain in 25% of patients and often, acute renal failure (Poutanen). Furthermore, MERS-CoV patients have a higher need for mechanical ventilation, with 50 - 89% (Paules et al.). Laboratory findings of MERS patients often include thrombocytopenia (low platelet count) and lymphocytopenia (low white blood cell count) (Poutanen).

The terminal events leading to death have been severe respiratory failure, multiple organ failure, sepsis or other complications such as acute myocardial infarctions. Post-mortem autopsies have revealed high viral RNA levels in lungs, bowels, lymph nodes and at lower levels in the spleen, liver and kidneys (Peiris et al.).

COVID-19

Phylogenetically, SARS-CoV-2 is much more similar to SARS than MERS, allowing them to share similar clinical manifestations (Petrosillo et al,).

The incubation period for COVID-19 could be up to 14 days, meaning that a person infected with COVID-19 could potentially live 14 days symptom-free, unknowingly carrying the virus. The median time until the onset of symptoms being 5.1 days (Lauer et al.). The common symptoms of SARS-CoV-2 are fever, cough, fatigue, and myalgia (Xu et al.). Less prevalent symptoms include gastrointestinal problems such as nausea, vomiting and diarrhea, loss of smell and taste and in younger patients, an itchy and painful skin condition that has been dubbed "COVID toes" (Marshall). Overall, 10% of patients require ventilation and ICU admission (Ragab et al.).

Just like SARS-CoV, SARS-CoV-2 binds to the ACE2 receptors within the body, which are extremely abundant in the lower respiratory system. In fact, COVID-19's affinity for the ACE2 receptor is even higher than that of SARS' (Petrosillo et al.). As the virus invades the cells lining the alveoli, the immune system sends white blood cells to kill the virus leaving a jumble of fluid and dead cells (pus) in the lungs. This physiological process interferes with regular oxygen exchange within the alveoli, creating shortness of breath and hypoxemia. This is when more intense symptoms begin to appear such as pneumonia and patients require mechanical ventilation (Wadman et al.). Conditions can worsen even further develop a condition called acute respiratory distress syndrome or ARDS which is marked by severe shortness of breath, dangerously low oxygen saturation, and low blood pressure and in most cases, results in death (Ratini). About 15% of cases develop ARDS (Ragab et al.).

Furthermore, due to an abundant amount of ACE2 receptors in the kidneys, it appears that acute kidney failure is another major problem that COVID-19 presents. 59% of nearly 200 hospitalized COVID-19 patients in China had proteinuria (protein in urine), and 44% had hematuria (blood in urine). The presence of both blood and protein in urine both indicate kidney damage (Wadman et al.).

More recently, there has been concern that COVID-19 causes a cytokine storm. Cytokine storm describes an event where the immune system becomes overly active. The body sends out too many cytokines, which are substances that are released as an immune response, and this triggers an aggressive reaction against the host's cells. Typically, cytokines are an important part of the body's immune response, however, in cases where the body becomes overwhelmed with pathogens, this condition may develop. If left untreated, this condition is quite deadly and can lead to multiorgan system failure and even death. Cytokine storms have been discovered in various viral infections such as influenza and even MERS and SARS (Ragab et al.).

Moreover, blood clots are another vital element that COVID-19 brings. It seems that COVID-19 causes major blood clotting, with one report claiming over 38% of patients had unusual blood clotting upon hospital admission. These clots are particularly dangerous as they can block major arteries within the heart and the lungs, an event that has killed a number of COVID-19 patients (Wadman et al.).

Lastly, a less talked about effect of COVID-19 is its ability to cause neurological damage. ACE2 receptors are present in the cortex and brain stem. Some COVID-19 patients have developed meningitis, encephalitis   andseizures   suggesting   that   COVID-19 can   penetrate   the   central   nervous   system   (Mao   et   al.).

Conclusion

As much as the viruses within the Coronaviridae family are similar in structure and function, as much are their diverse differences in clinical manifestations. Coronaviruses can be as mild as a common cold or severe enough to cause death after two weeks of infection. These viruses' effects are often not localized, with problems ranging from a simple upper respiratory infection to encephilits to even acute kidney failure. Day by day, scientists and researchers work hard to uncover the mechanisms by which these viruses devastate the human body. Understanding the symptoms of older coronaviruses may help us better predict the symptoms and pathophysiology of COVID-19, prevent unnecessary crises and speed up developing effective treatments. Overall, understanding the real consequences of COVID-19 infections should and will shape our decisions as part of a larger global community moving forward.

CHAPTER 4
SYMPTOMS OF CORONAVIRUS INFECTIONS IN ANIMALS

BY MIRAY MAHER

Introduction

As the world tries to navigate around this novel coronavirus, many questions arise. What kind of conditions does this virus bring about? What tissues does it target? What are the long-term health effects? However, a whole new dimension of complexity is introduced when it is factored in that certain animals are capable of being infected by SARS-CoV-2 and acting as a transmission link. Coronaviruses naturally circulate among a diverse array of animals, causing respiratory tract infections in some species and gastrointestinal in others. Identifying the symptoms of COVID-19 infections in wildlife and domesticated animals will help us eliminate further outbreaks caused by animal to human transmission and to control the pandemic. Furthermore, this chapter will explore the different strains of coronaviruses that infect the domestic and wild animals that directly or indirectly affect our day to day life.

SARS-CoV-2 and Animals

Panic ensued at a mink farm in the Netherlands when two dozen minks became infected with the novel coronavirus, SARS-CoV-2. This is a case of reverse zoonosis, or, a viral spillover infection in the "opposite direction" meaning that the minks caught this potentially deadly virus from an infected farm handler. These minks have since developed a runny nose, difficulty breaking, gastrointestinal problems and even pneumonia leading to death. In addition, the infected minks have shown that they could spread the virus back to humans, contributing to the COVID-19 pandemic. Thus, the Dutch government decided to cull all the minks on the farm to prevent the problem from exacerbating (Enserink).

This problem sparked a huge discussion on which animals are capable of catching the virus and spreading it either to members of their own species or across species. Researchers have begun to conduct studies to understand the infection risks that animals pose to humans and vice versa amidst this pandemic.

The animals most in contact with humans are pets, the majority of which are cats and dogs. According to the CDC, there has been a limited number of animals infected with SARS-CoV-2 and they likely do not play a major role in the propagation of this pandemic. In a study, cats were successfully inoculated with the SARS-CoV-2 virus and examined at regular intervals. The results show that cats are highly susceptible to the virus, as upon necropsy found large lesions in the mucous membranes of the nose and trachea and in the lungs. Furthermore, the cats transmitted the virus to other cats via airborne respiratory droplets. In addition, younger cats had more severe symptoms than older cats (Shi et al.). It is still unknown whether cats could transmit the virus to humans. It is also worthy to note that these results were obtained in a field setting and very few cases were reported within a domestic setting (Public Health Agency of Canada). By the same token, other felines like tigers and lions can also catch SARS-CoV-2. On April 5 of this year, the Bronx zoo announced that some of the tigers and lions at the zoo exhibited a dry cough, lack of appetite and other mild symptoms. They have since made a full recovery. They were likely infected by an infected but asymptomatic zookeeper, however, there is no evidence to suggest that felines can spread the novel coronavirus to humans (Daly).

Dogs, on the other hand, are much harder to infect with the novel coronavirus. Experiments on beagles inoculated with the virus show that although viral RNA was detectable in a rectal swab in only some dogs, there was no viral RNA in any organs or tissues. The dogs did not exhibit any symptoms (Shi et al.). This shows that dogs have a very low susceptibility to COVID-19. Around the world, very few dogs have tested positive for COVID-19 with one being a Pomeranian in Hong Kong (Daly).

Ferrets, which share the mustelid family with minks, are also susceptible to SARS-CoV-2. The virus replicated in the upper respiratory tract for 8 days without causing severe symptoms or death (Shi et al.).

Hamsters, another popular family pet, can be infected by the novel coronavirus and can spread it to other members of the same species. Mice, although another member of the rodent family,

34

cannot be infected by this virus (Public Health Agency of Canada).

Livestock is another sector that has been greatly focused on as outbreaks in an agricultural setting could be disastrous in terms of food security and the economy. Fortunately, pigs, chickens and ducks have all been shown to not be susceptible to SARS-CoV-2 (Shi et al.).

Furthermore, non-human primates, especially apes, are suspected of being capable of getting sick with this new virus. The target receptor of COVID-19 is the angiotensin-converting enzyme 2 or ACE2, which is identical to its human counterpart, raising suspicions about the infection risk in apes like gorillas (" Protecting Great Apes from Covid-19"). In China, scientists have successfully infected rhesus macaques, a type of monkey. They developed pneumonia and the virus proliferated in the respiratory and gastrointestinal tracts, with recovery occurring after approximately two weeks. More importantly, the scientists discovered that the monkeys developed immunity to the virus as when they attempted to re-infect the monkeys, they only developed a mild fever but no other symptoms (Deng et al.). Studying the antibodies developed in the rhesus macaques, whose DNA is 93% similar to humans (Choi), could help build the framework by which vaccines could be developed and gives clues as to how the human body reacts to COVID-19 re-infection.

Bovine Coronaviruses

The bovine coronavirus, or BCoV for short, infects cows all over the world and devastates the agricultural sector. The bovine coronavirus is a pneumoenteric virus meaning that it causes respiratory and gastrointestinal symptoms. Specifically, it affects the upper and lower respiratory tract and the intestines in cows (Saif). The virus spreads through respiratory droplets and through the feces (Oma et al.).

Bovine coronaviruses cause three distinct diseases in cows including calf diarrhea, winter dysentery in adult cows, and respiratory illness which afflicts cows of all ages and results in shipping fever or bovine respiratory disease complex (Saif).

Calf diarrhea, or scours, a disease caused by a strain of coronavirus that affects newborn, unweaned calves. A calf develops immunity to viruses as such by drinking its mother's milk. The first milk produced after giving birth is called the colostrum and is rich in maternal antibodies. Because a neonatal calf still has not developed any immunity, calf diarrhea can be fatal, with a reported mortality rate of about 50% (Potter). Clinically, sick calves have watery, yellow diarrhea that may contain blood or mucus, dehydration, fatigue, and poor growth. If treatment is not promptly administered in terms of rehydration and electrolyte replenishment, death may occur as soon as 24 hours after infection. Depending on the severity, calf scours can last from a couple of days to up to two weeks (Gould).

Unlike calf diarrhea, winter dysentery affects adult cows. As the name of the condition suggests, winter dysentery is most prevalent during the winter months. It mainly affects housed dairy cows between the ages of 6-24 months. (Gruenberg). It is highly contagious and is transmitted through the fecal-oral route, however, airborne respiratory droplets are another plausible transmission pathway (Marker). Symptoms include explosive diarrhea that is dark green or black in colour and may have blood, intestinal inflammation with lesions, drop in milk production (up to 95%) and mild respiratory symptoms such as coughing and nasal discharge (Gruenberg; Marker). Although cows afflicted with this condition will feel very sick, mortality is relatively low (1-2%) and recovery can be expected within a week (Gruenberg).

In general, bovine respiratory disease complex or BRD refers to any disease affecting the lower and/or upper respiratory tracts in cows. Clinical features include fever, dyspnea, coughing, nasal discharge, poor appetite and lethargy. It is caused by multiple factors interplaying in a way conducive to this illness. Firstly, certain characteristics make a host more susceptible to this disease than others like old age, poor immunity and genetics. Secondly, poor living or environmental conditions like crowding or poor ventilation could be another contributing agent. Lastly, the cow must be exposed to the harmful pathogens that cause this disease which include viruses like the bovine coronavirus, bacteria and parasites. Oftentimes, ill cows are infected with a virus that weakens them and then are exposed to bacteria that leads to severe pneumonia. Unfortunately, this disease is the leading cause of death among cows and has a high mortality rate of 45-75%, devastating many in the agricultural industry (Janzen and Jelinski).

Feline Enteric Coronavirus

Prevalent in domestic cats, the feline enteric coronavirus causes very mild gastrointestinal symptoms of diarrhea and vomiting but often resolves without treatment. Some cases have also reported mild upper respiratory symptoms. This virus is spread through the ingestion or inhalation of virus-contaminated feces (Gallagher). 50% of cats in single cat households will get infected with the coronavirus and up to 90% in multicat households, making this virus very common. However, sometimes, this mild condition may develop into something more deadly called feline infectious peritonitis (FIP). About 5-10% of FCoV cases develop into FIP. First symptoms may be lethargy, decreased appetite, and a fever. After a couple of weeks, cats may develop a 'wet' form of FIP which involves fluid accumulating in the body's cavities like in the abdomen or in the chest, leading to dyspnea. Other cats may develop a 'dry' form of FIP which does not involve any fluid buildup but the severe inflammation of organs such as the brain, liver and the intestine. Sadly, this disease will result in death in most cases (Hunter).

Canine Coronaviruses

The canine coronavirus or CCoV is a highly infectious disease that causes intestinal infection in dogs. Transmission occurs through oral contact with contaminated fecal matter (Gollakner). The CCoV symptoms often present in the enteric tract but viral RNA has been isolated from all tissue except the brain. Necropsy of affected dogs showed severe lesions in the intestinal tract, spleen, liver, kidneys and lymph nodes.

This disease is more fatal to puppies than in older dogs, with younger puppies even exhibiting neurological signs like seizures and ataxia

(Buonavoglia). The symptoms of this infection resemble that of enteritis, however, mixed infections often result in a more severe course of illness. Clinical symptoms include a fever, sudden onset of diarrhea with an orange tinge and a putrid odour that can sometimes contain blood, decreased appetite and lethargy (Gollakner).

Porcine Coronaviruses

The three strains of coronavirus afflicting swine enterically are the transmissible gastroenteritis virus (TGEV), porcine epidemic diarrhea virus (PEDV), and porcine deltacoronavirus (PDCoV). A virus similar to TGEV is the porcine respiratory coronavirus (PRCV), which is the result of a mutation in the spike proteins. TGEV first appeared in 1946 and PRCV in Belgium in 1984. Since, these viruses have rapidly spread to the point that most pigs are immune to them and are no longer a source of economic loss. On the other hand, PEDV and PDCoV, which have emerged more recently in 1971 and 2009 respectively, still pose a major threat to pigs, especially piglets (Valsova et al.).

Porcine epidemic diarrhea virus is a highly infectious disease that targets the small intestines of pigs, killing cells within the intestinal tract (Koonpaew). This results in symptoms of acute gastroenteritis like profuse watery diarrhea and vomiting that can lead to severe dehydration especially in neonatal piglets. In fact, this disease has an 80-100% morbidity and a 50-90% fatality in suckling piglets (Koonpaew). In 2013, PEDV devastated the United States and also affected Canada and Mexico, causing the deaths of more than eight million piglets in the United States alone in one year. In addition, rampant PEDV outbreaks occurred in Asian countries like South Korea, Taiwan, and Japan. All in all, PEDV is one of the most devastating infectious diseases to hit the global pork industry (Lee).

The novel porcine deltacoronavirus, PDCoV, causes similar symptoms to PEDV however, its mortality rates are reportedly lower and the only fatalities have occurred in suckling pigs. Both viruses are transmitted through the oral-fecal route (Hu et al.; Lee).

Conclusion

Coronaviruses, which are zoonotic by nature, greatly affect animals and humans alike. The novel coronavirus has overcome interspecies barriers from bats to what is believed to be a pangolin to lastly humans (Fox). However, the transmission pathway does not stop there, as other species such as cats and ferrets can also catch this virus and spread it to other species, members of their own species and even back to humans. Furthermore, understanding the viruses that naturally affect animals around us can help us better predict the mutations necessary for them to be able to successfully infect humans. Therefore, understanding animal coronaviruses is crucial to understanding the pathophysiology behind human coronaviruses as human coronaviruses have evolved from them. Overall, understanding the coronaviruses and diseases that afflict animals will help humans establish the adequate ecological boundaries to prevent the spread of these diseases to other species, and especially to the human population.

CHAPTER 5
EVOLUTION OF CORONAVIRUSES IN THE ANIMAL KINGDOM ACROSS HISTORY

BY TINA WU

From the time that SARS-CoV-2 first surfaced at the Huanan seafood market in Wuhan, China, the origin of the virus has been hotly debated, speculated upon, and researched. Coronaviruses are zoonotic. These viruses originate from animal hosts but occasionally come across a rare mutation that allows them to jump to another species. When the first outbreak of COVID-19 occurred in December 2019, evolution was at play, causing the virus to jump from animals to humans. This shift in the virus's ecological niche is a critical point for the start of the COVID-19 pandemic. However, the story of SARS-CoV-2 does not begin with its transmission in the Wuhan wet markets. Rather, its tale begins tens of thousands, or perhaps even millions of years ago.

Mechanisms of Viral Evolution

Before exploring the evolutionary history of coronaviruses, the basic mechanics of viral evolution must be understood. Viral evolution refers to the heritable genetic changes that a virus accumulates in its lifetime, which can arise from environmental adaptations, or adaptations in response to host immune systems. Viruses evolve rapidly due to their short generation times and large population sizes ("Viral Evolution"). Traits that increase a virus's ability to reproduce, such as high infectivity, will become more prevalent as time progresses through natural selection.

Natural selection, a theory first proposed by Charles Darwin and Alfred Russel Wallace in 1859, is a process by which the genetic characteristics that increase reproductive fitness are passed down from generation to generation. Natural selection has three criteria ("Natural Selection"):

1. *Differential reproduction must occur.*
2. *Natural selection requires that the traits are heritable.*
3. *There must be genetic variation.*

The first criterion states that certain genetic characteristics must be better for reproducing, while others are worse. In the context of viruses, certain viruses may have characteristics that make them more or less infective.

The second criterion implies that the viral traits must have a genetic basis in the form of DNA or RNA that can be passed down to offspring. Deoxyribonucleic acid (DNA) and ribonucleic acid (RNA) are crucial to life, as they encode the information for how the cell or virus is constructed and how it can be replicated. This information is coded in the form of a sequence of nucleotides that contain instructions on the production of vital proteins. In DNA, the nucleotides are adenine (A), guanine (G), cytosine (C), and thymine (T). In RNA, the thymine is replaced with uracil (U). While coronaviruses consist of RNA, there are viruses that encompass DNA instead.

The final criterion of natural selection is on genetic variation, which, in viruses, is achieved in two main ways: mutation and recombination. Mutations occur due to copying mistakes in genomic replication; they are not deliberate changes but random incidents, as no virus can decide to become more or less infective. Viruses tend to be far sloppier with replication than the cells in our bodies, which leads to a higher mutation rate. In human cells, there are specialized enzymes that proofread copied genetic information to ensure there are no mistakes, but many viruses lack this function causing them to be more prone to copying errors ("Virus Strains"). Common mutations include substitutions, insertions, and deletions. Substitutions are the replacement of one nucleotide in a sequence with another. For instance, when we compare TACATC and GACATC, we can see that the first nucleotide changed from a thymine to a guanine. Insertions occur when a nucleotide is added to a sequence, while deletions occur when a nucleotide is removed. If we consider a sequence such as AGACUA, and delete a single nucleotide: AGCUA, we can see that all the letters after G have shifted to the left by one space. This is why insertions and deletions are also known as frameshift mutations (DiGiuseppe and Adam-Carr 340).

The second way viruses achieve genetic variation, recombination, typically occurs when two viruses infect the same cell at the same time. When a virus infects a cell, it essentially hijacks the cell machinery to produce more copies of itself. During this process, a large quantity of genetic material is floating within the cell, and recombination is more likely to occur. One form of recombination is when the pieces of DNA or

RNA physically break and reconnect as similar sections of viral genomes pair up and swap pieces. The second form of recombination is called reassortment, which occurs when whole segments of viral genetic material are swapped ("Evolution of Viruses (Article) | Viruses"). Viral segments are similar to chromosomes in humans; the genome is split between a number of self-contained blocks. Reassortment is the exchange of these segments. While highly relevant with flu viruses, this type of recombination does not occur in coronaviruses, which have unsegmented genomes.

The Coronavirus Common Ancestor

The current understanding of the evolutionary history of coronaviruses is a mosaic of modern-day observations, models, and genomic sequencing techniques. Researchers try to piece together the puzzle of the past as best they can, but oftentimes there are entire missing chunks or areas that simply do not fit together quite right. Without a time machine to go back hundreds or thousands of years to examine the viral life on Earth, it can be difficult to understand how or when the ancestor to coronaviruses first emerged, and how it evolved into the viruses seen today. The next best possibility for scientists is to examine the present-day viruses and make inferences on their evolutionary past.

We start at the beginning. All varieties of coronaviruses today likely originated from a single common ancestor that existed many years ago. Molecular clock dating analyses suggest that the most recent common ancestor to present-day coronaviruses existed 10,000 years ago. This value is calculated by estimating that the viral genetic sequences had a mutation rate of 1.3×10^{-4} nucleotide substitutions per site per year. Unfortunately, there are issues with this analysis. It is understood that RNA viruses can mutate quickly because they lack the proofreading mechanisms our cells have to ensure the genetic material was copied correctly. However, coronaviruses have a special proofreading mechanism of their own. Coronaviruses, in fact, have slower mutation rates closer to those of DNA viruses. If we keep the number of mutations the same, but decrease the mutation rate, it takes a longer amount of time to reach that same quantity of mutations. Therefore, the length of coronavirus evolutionary history may be greatly underestimated and a case can be made for a more ancient evolutionary past of the coronavirus (Wertheim et al.).

Another piece of evidence that could support the idea of an ancient common ancestor considers the natural host species of coronaviruses. During the SARS epidemic of 2003, researchers were prompted to examine sources of coronaviruses in humans, domesticated animals, and wildlife. They found that the greatest diversity and abundance of coronaviruses were found in bats and birds, pointing to these as its natural hosts. This is consistent with the hypothesis of an ancient common ancestor. Birds and bats evolved tens of millions of years ago. If the coronavirus was truly only 10,000 years old, this would be incredibly young compared to the hosts it infects (Wertheim et al.).

A final plausible explanation is purifying selection (Wertheim et al.). As mentioned, mutations are completely random in nature, and no virus can decide to gain or lose certain characteristics for the purpose of furthering its progeny. In fact, most mutations are deleterious. Natural selection has done a phenomenal job of creating and promoting the structures best suited to viral survival, and any mutation that strays from that is more likely to be harmful than beneficial. Therefore, purifying selection or negative selection maintains the long term stability of biological structures by protecting against unfavourable mutations (Loewe). Strong purifying selection could hide ancient evolutionary histories. In theory, it could slow down the rate of evolution by an order of magnitude relative to the rate of silent or synonymous mutations, which are mutations that do not affect gene functioning (Wertheim and Pond). The mutation rate is calculated by the number of mutations divided by the time it takes for those mutations to arise. If the rate stays the same, but silent mutations are accounted for, there will be an increase in the total number of mutations that occurred. It must then follow that the time it takes for these mutations to come about also increases. This implies that the time since the surfacing of the most recent common ancestor of the coronavirus is underestimated; the coronavirus might be a much far more ancient than molecular dating evidence would suggest.

Bioinformatics in the Discovery of Evolutionary Lineage and Relationships between Coronaviruses

Tens of thousands, or perhaps millions of years ago, a virus was born that mutated and diversified into all the types of coronaviruses in existence today. Currently, the coronavirus family is split into four groups or genera: alpha, beta, gamma, and delta coronaviruses. Alpha and beta coronaviruses infect mammals, while gamma and delta coronaviruses primarily infect avian species (Cui et al.). The path of divergence of these genera is not clear-cut. Scientists must examine the genetic similarities between present-day coronaviruses to make hypotheses about the order in which viral categories branched off in the past. Such an analysis requires the use of bioinformatics. This field examines the similarities and similarities and differences in protein or genetic sequences. Bioinformatics combines the fields of computer science, biology, information technology, and mathematics to solve complex biological problems, such as the relatedness between coronavirus species.

An algorithm is a series of step by step processes that one applies to solve a problem. The algorithm for cake baking is the recipe. The algorithm for finding your way to an ice cream parlor is a set of directions. The algorithms for determining coronavirus relatedness are mathematical equations. A traditional method that can be used to determine the similarities between various coronavirus genomes is sequence alignment. In its simplest form, sequence alignments involve comparing genetic sequences of the same length side by side and counting the number of matching nucleotides. The similarity between the sequences is known as the alignment score and the dissimilarity is known as the distance (Prjibelski et al.). If viruses have a high alignment score, they are likely more closely related. If the viruses have a low alignment score, they likely diverged from each other a longer time ago. Algorithms are developed based on this idea to accurately calculate the similarities between two sequences. Sadly, these algorithms require a lot of computing power—they use too much memory and have too high of a run-time to be applicable in real life (Prjibelski et al.). This is especially true when examining large gene sequences or comparing a sequence with many others. Therefore, other methods are developed.

A heuristic algorithm in computer science is a technique used to solve a problem more quickly, but sacrifices optimality, precision, or accuracy for speed (Kenny et al.). Basic Local Alignment Search Tool (BLAST) is a widely used heuristic search algorithm that improves the speed of sequence comparisons. BLAST first takes the genomic sequence and divides it into smaller components called seeds. These seeds are then searched throughout the entire database for exact matches. When a database sequence matches a seed exactly, it is expanded upon to create the longest matching sequence possible. BLAST will miss a small portion of significant matches, but overall, it is one of the most widely used search algorithms in the biological world due to its accuracy and sensitivity (Saeed et al.). Other methods of genomic comparisons include machine learning using neural networks or hidden Markov models. In papers specific to coronavirus phylogeny, there have been applications of the Poisson model, while another uses a geometric approach (Zheng et al.).

As algorithms are improved over time, the understanding of evolutionary lineage changes. With our current knowledge, it remains unknown at what point and in which order the four different genera branched away from each other and diversified. There are many hypotheses, but no definitive answer. However, a visual of the diversification of coronaviruses can still be formed. A phylogenetic tree is a depiction of the relatedness between species using branches that show the diversification of a species from a common ancestor. But rather than constructing phylogenetic trees in the standard linear timeline format, many papers on coronavirus phylogeny will create one in a radial format, which allows for the visualization of the relatedness between coronavirus species without identifying the root.

Conclusion

Just as coronaviruses are constantly evolving, so is our understanding of their evolutionary lineage. Molecular dating techniques suggest the most recent common ancestor to coronaviruses is 10,000 years old. One paper attempts to dissuade this finding by arguing for an ancient evolutionary past of coronaviruses. This newer paper might even be proven incorrect as more research surfaces. The process of diversification of the four coronavirus genera from this common ancestor is also admittedly unknown, with different results depending on the sequencing algorithm used. This uncertainty manifests itself in the use of radial format phylogenetic trees rather than typical linear formats that imply a timeline and sequence of events. While the scientific field contains a wealth of knowledge, scientific progress itself is the result of admitting ignorance rather than pride. The truth of the world is only the truth of the moment; as progress is made, foundational understandings will shift. This humility allows humankind to continue its curious pursuit for answers about the world's functionings and slowly take steps towards the truth, as more knowledge surfaces each and every day.

CHAPTER 6
ENVIRONMENTAL FACTORS IMPACTING COVID-19 SPREAD

BY MIRAY MAHER

Introduction

COVID-19 has taken the world by storm. It has left many countries in economic turmoil with many people sick and unemployed. Many have kept up to date on the number of global cases on Worldometers, and watched as they drastically grew from only 12,000 cases in February, to almost 14,000,000 in July. This raised many questions within the scientific community about how the virus quickly spread, its incubation period and how long it can survive without its host. Moreover, as the whole world anticipated a vaccine, people speculated that the pandemic would fade as summer came along and the temperatures rose. However, is this really the case? Will the COVID-19 pandemic be gone by the end of the summer? This chapter will attempt to answer this pressing question by examining the weather parameters that promote COVID-19 spread. This chapter will cover different environmental and weather conditions and how they may affect the spread of COVID-19, including: air quality, temperature, humidity, precipitation, wind speed, climate change, and land transformations. It's worthy to note that although the following information pertains to how weather conditions impact transmission, similar results apply to the mortality rate (Yueling et al.).

Air Quality

Firstly, let's examine how air quality aids or hinders COVID-19 spread. Studies have shown that air pollution is linked to the prevalence of viral diseases like influenza. The air contains particulate matter, which is the sum of all solid and liquid particles suspended in the air. Many of these particles are hazardous and may remain suspended in air for a very long time. Viral particles can remain in the air longer when they are suspended on the particulate matter. These viral particles can be deeply inhaled within the respiratory tract where the virus can penetrate the lung's epithelial cells and move to other organs, inducing an infection. Particulate matter that is small enough will remain airborne for even longer periods of time due to low settling velocity will have a greater capacity to infect people (Xu et al.). Thus, poor air quality can definitely contribute to a faster SARS-CoV-19 spread.

Temperature

Temperature is another factor that has been explored greatly. Lower temperatures have helped with transmitting viruses such as SARS and MERS and it is speculated that higher temperatures might kill off the COVID-19 virus, like it did with the 2003 SARS virus in Guangdong, China (Yueling et. al). In fact, most new cases occur globally between 3 °C and 17 °C (Gupta et. al). Before March 11, 90% of the SARS-CoV-19 cases were recorded in places with temperatures less than 11° C (Bukhtari and Jameel). This shows that temperature most likely plays a critical role in viral transmission, with higher temperatures slowing down its spread. Unfortunately, exhaled COVID-19 viral particles are covered in saliva and mucus which makes them resistant to extreme weather conditions. Thermal defense mechanisms also exist within humans and other mammals as they have evolved to have elevated body temperatures that also prevent viruses and other pathogens from surviving within their bodies. The standard body temperature of 37°C is fatal for many microbes and pathogens. Therefore, the human body is a natural thermal barrier to these harmful organisms. However, as global warming continues, natural selection begins to favour organisms that are naturally more suited to warmer climates. This could prove detrimental to humans as our natural thermal barrier becomes futile (Gourdazi). In addition, increasing global temperatures may impair the normal function of an immune system. Research at the University of Tokyo shows that elevated temperatures may compromise with the immune system's response to pathogens. This conclusion was attained by infecting three healthy mice with an influenza A virus (causes the flu in humans) but placing them in 3 different temperature-controlled spaces, being 4°C, 22°C and 36°C. The mouse in the hotter temperature did not fight off the virus as well as the other, cooler mice, perhaps due to the higher thermal energy altering the activity of specific genes. It was also noted that the mouse in the hottest environment did not eat as robustly as the other mice. This contributed to a potential nutritional deficit that could have played a part in the weaker antiviral immune response (Moriyama and Ichinohe). From this study, it could be extrapolated that similar physiological effects may be apparent in humans in hotter temperatures, however, it is unclear if it is because of a direct link between temperature and the immune system or the greater issue of global food security that is hampered by the quickly changing environmental conditions. What this means is that a possible reason for the rate of transmission of the SARS-CoV-2 virus could be the rapid rising of global temperatures (Gourdazi).

Humidity, Precipitation and Wind Speed

As previously mentioned, viruses within the Coronaviridae family thrive in lower humidity. Higher humidity causes the saliva and mucus covered viral particles to settle very quickly, preventing them from spreading as effectively. According to Gupta et al., absolute humidity is a better metric to study the relationships between COVID-19 and weather conditions. Absolute humidity is a measurement that is expressed in g/m3. It describes the actual mass of water per measure of air volume. So if the absolute humidity of a geographic area is 50 g/m3, that means that there is 50 grams of water vapour in 1 m3 of air. On the other hand, relative humidity is expressed as a percentage. It is the ratio between the amount of water in the air and the maximum amount it could hold in the given temperature and pressure (Khillar). Most new cases are occurring at an absolute humidity between 4 and 9 g/m3 globally (Gupta et al.). These are relatively low values. Many airborne viruses, such as the COVID-19 virus, are sensitive to ambient humidity. The high air water content can cause the viral structure to become damaged through the removal of water molecules from the capsid (structure surrounding viral RNA). Moreover, due to the hydroponic nature of the lipid membrane enveloping the virus, there is stress induced through the water molecules pulling on the virus and increased surface tension causing the viral molecules to break, tear or change shape (Xu et al.). Precipitation and wind speed had no correlation with the doubling time of the number of COVID-19 cases (Bukhari and Jameel)

Respiratory Illnesses: Winter vs. Summer Months

Multiple viruses from the Coronaviridae family like SARS CoV and MERS-CoV have demonstrated a seasonal preference and thrive in low temperatures and humidity (Gupta et. al). Due to the low humidity and lower temperatures that accompany late winter and early spring, most respiratory diseases are common during those seasons (Xu et al.). While it might be anticipated that the warm summer months will change the prevalence of disease, it turns out that certain aspects of hot temperatures mimic the conditions of the cooler months. Hot temperatures force people to stay indoors more often and turn on air conditioning, paralleling the drive to stay indoors during the bitter winters. Unfortunately, this causes respiratory illnesses to rise as people indoors tend to rebreathe the same circulating air that is little refreshed from the outside (Powell). This partly explains why certain Southern American states, like Florida, Texas, and Arizona, are experiencing high numbers of daily new COVID-19 cases despite their hot weather. In Wuhan, hospitals were asked to keep windows open 24 hours a day with no air conditioning in inpatient wards to prevent patients from rebreathing recycled air and contributing further COVID-19 spread (Yueling et al.) Therefore, although the virus survives less in hotter weather, the extreme heat causes people to go indoors, close windows, reducing ventilation by turning on AC. As global temperatures continue to rise, perhaps there will be a surge in respiratory diseases due to longer periods spent indoors.

Climate Change and Virus Transmission

Although there is no evidence that climate change is directly affecting COVID-19 spread, global warming may cause animals to migrate to cooler areas such as the poles that usually would not do so. On average, land species shifted 17 kilometers towards the poles every decade, while marine animals moved 72 kilometers per decade. (Pecl et al.) This rearrangement of species around the planet causes animals that host unique pathogenic microorganisms to meet other species that naturally would not get into contact with each other, spreading disease across species and creating novel transmission pathways.

Land Transformations and Agriculture

Furthermore, as humans continue to transform natural habitats into agricultural land, viruses that transfer from animals to humans, such as the COVID-19, will become even more common. The transformation of forests into farmland has left many people living on the edge of natural habitats, increasing their chances of contracting pathogens from disease-harbouring wild animals. An example of this is HIV, which jumped from primates to humans through close contact with infected bodily fluids. Thus, as humans continue to invade the space of these animals, the boundaries are blurred between wild habitats and human civilization, allowing the diseases to migrate from species to another (Stanford University). Moreover, habitat loss causes animals to migrate to non-endemic areas, causing them to cohabitate a space with animals that host pestilent microorganisms that can spread from species to species (Kenney et al.) When a pathogen that is prevalent in one species transfers to another species, this is described as virus spillover and will be discussed in further detail down the line (Kenney et. al). Other things that contribute to this viral spillover is the livestock industry. The demand for meat products combined with food insecurity may lead farmers to undertake unsuitable and often harmful husbandry practices that lead to infection spreading across species. The COVID-19 virus is theorized to have originated in this way through wet markets. Wet markets, as described in an article by National Geographic are "typically large collections of open-air stalls selling fresh seafood, meat, fruits, and vegetables." Live animals may be slaughtered on site and the selling of wild animals, such as bats, rats and snakes (Maron). These wet markets are often the breeding ground diseases and spillover infections due to unsanitary conditions. Crates placed on top of crates allows fluids such as feces, blood, saliva and urine from animals above to trickle down to the confined animals below. This allows pathogens from one animal species the opportunity to mutate and penetrate the cells of another species. The jumble of bodily fluids act as a perfect mediator for the spread of these diseases (Bernstein and Salas.)

How Spillover Events Occur

As previously mentioned, spillover infections describe the event when a virus overcomes the natural biological barriers that prevent a species from infecting another. There's five-step series of events that outlines the series of events leading up to a spillover infection (Kenny et al.):

1. *Reservoir:* The virus proliferates within the host. It is crucial that the virus not be too pestilent, so that the host is able to live to past it on. In this way, the host species are "holding" the virus or acting as a reservoir for it (Rydling).
2. *Exposure:* The secondary species comes into contact with the primary host species, causing them to be exposed to the virus (Kenny et al.).
3. *Breakthrough:* The virus is able to overcome natural occurring barriers such as incompatibility with the new species and the immune response of the secondary host (Kenny et al.).
4. *Transmission:* The virus spreads from one new host to another (Kenny et al.)
5. *Tipping point:* The virus effectively spreads within the new species; the virus transmits efficiently and rapidly (Kenny et al.)

Coronaviruses (the family to which SARS-CoV-2 belongs to) are zoonotic in nature (originating in animals and capable of spreading to humans). Spillovers exist all around us, with the novel coronavirus being a prime example. The common flu is caused by the influenza virus which originated in birds and spilled over to humans (Rajewski). Similarly, the Ebola outbreak in 2016 was traced back to a single infected person in the Democratic Republic of Congo who caught the disease from a bat (Ries). Unfortunately, spillover infections could also indirectly harm humans by impacting the food supply. Spillover events could occur animal to animal, creating illness within livestock. A prime example of that is the porcine epidemic diarrhea virus (PEDv), which originated in bats and has been transmitted to pigs. This virus is very deadly, with mortality rates ranging from 10%-100% in piglets (Kenny et al.). This porcine coronavirus devastated the pork industry, killing around eight million pigs in 2014 (Sung et al.).

Overall, all these virulent illnesses affecting people were caused through close contact with disease-harbouring wild-life, whether by consumption, inhaling respiratory droplets, or otherwise sharing the same space. Most people would agree that they would avoid any risky contact with wildlife, however, there are various compelling socioeconomic factors that might force someone to rely on resources existing near these potentially pestilent animals or even directly involving them (ex. consumption). These causes may include food insecurity and demand for animal protein, urbanization, climate change and lack of fertile agricultural land. Studying the root causes behind these spillover events can help prevent them from occurring in the future.

Conclusion

In conclusion, exploring the effects that different weather conditions have on COVID-19's spread is essential in identifying vulnerable geographic regions, which can help governments take the necessary action to preserve life and diminish losses. Additionally, identifying the behaviours that drive spillover infections, like land transformations, deforestation, and consuming wildlife and their root causes will prove valuable in preventing them. As the pandemic situation continues to evolve, these valuable trends will serve as a starting point for change and edification for the way future outbreaks are handled.

CHAPTER 7
VIRAL CHARACTERISTICS THAT CHANGE INFECTIVITY

BY TINA WU

Coronavirus disease 2019 (COVID-19) was officially declared a pandemic by the World Health Organization on March 11, 2020 (Ghebreyesus). As of August 2020, the highly infectious pathogen continues to ravage the globe and afflict millions of individuals worldwide. This outbreak began with a zoonotic transfer from animal hosts to human hosts. Since its introduction to the human population, the virus has caused epidemics in numerous countries while healthcare professionals, governments, and the general population combat its spread. Structural characteristics of the virus can be examined to reveal the biological causes of the high transmission rate of COVID-19. Hence the infectivity of the COVID-19-causing virus, severe acute respiratory syndrome coronavirus 2 (SARS-CoV-2), is explored in relation to its viral characteristics.

Introduction to Viral Characteristics

While the vast majority of viruses are too small to be seen with an ordinary light microscope at only 5 to 300 nanometers in size, their simple yet effective structures have allowed them to become mind-bogglingly abundant with million of times more viruses than stars in the galaxy (Kaiser; Wu). Viruses display a vast array of shapes and sizes, but they have certain basic traits in common. All viruses feature a nucleic acid genome of single- or double-stranded DNA or RNA tucked inside a protective protein shell known as a capsid. The capsid is coded for by the viral genome but due to the genome's limited size, only a few structural proteins can be coded. Therefore, capsids consist of only one or a few protein species which self assemble to form a continuous 3-dimensional structure in either a helical or icosahedral fashion. Some viruses, including coronaviruses, have an additional protective covering called an envelope: a lipid bilayer studded with glycoproteins that form spikes on viral surfaces (Gelderblom).

Background on the Viral Life Cycle

Viruses are sometimes described as undead—not in the way that zombies are undead, but because they are unable to reproduce on their own. They must rely on a host cell's machinery and resources to create viral progeny. As such, viruses are obligate intracellular parasites, neither classified as living nor non-living. The viral life cycle describes a series of steps in which a virus will enter the host cell, "reprogram" it, and use the host's resources to make more viral particles. While the replication life cycle can vary greatly from virus to virus, there are five basic steps ("Intro to Viruses (Article)."). The ability of the virus to complete these steps is crucial to its infective rate.

Viral Life Cycle

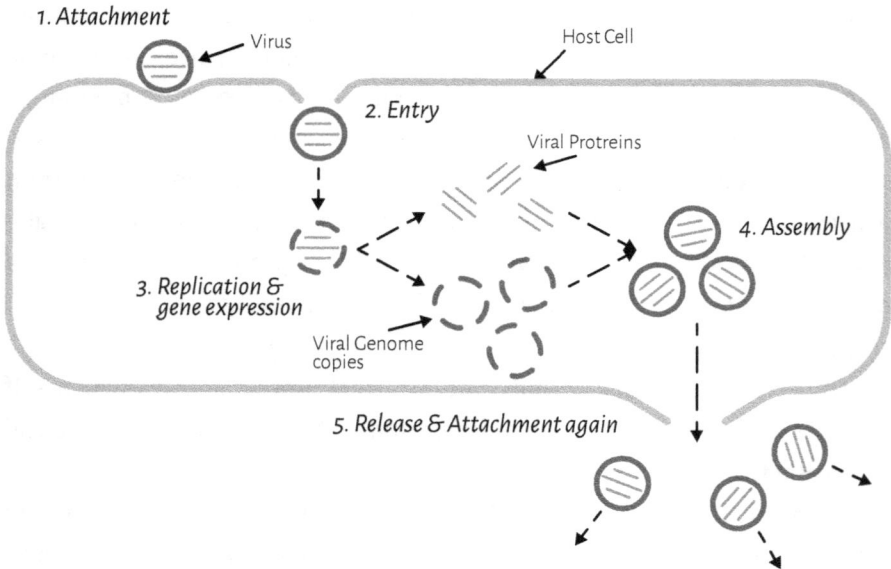

1. *Attachment*

Virus

Host Cell

2. *Entry*

Viral Protreins

4. *Assembly*

3. *Replication & gene expression*

Viral Genome copies

5. *Release & Attachment again*

1. **Attachment:** *viral proteins on the capsid or lipid envelope recognizes and binds to the host via specific receptors on the host cell surface. The specificity of the viral protein determines the varieties of cells, tissues, and organisms the virus can infect, known as its tropism.*
2. **Entry:** *the binding of the viral protein and the host cell surface receptor induces conformational changes allowing the virus or viral genetic*

material to enter the cell.

3. **Replication:** *the viral genome is copied and the genes are expressed, creating new viral proteins. The process of transcription occurs when the viral genetic material is used to create messenger RNA (mRNA), a necessary step in gene expression. The mRNA is then converted into proteins by ribosomes in a process called translation.*
4. **Assembly:** *new viruses are formed by packing together the newly synthesized viral proteins with a copy of the viral genome.*
5. **Release:** *the virions (completed viruses) are released from the cell (Goulding).*

The Role of Viral Characteristics in the Coronavirus Life Cycle

In all aspects of biology, whether it be the actions of the human body or those of pathogens, structure and function are inseparably linked. Changes in infectivity rely heavily on changes in viral structures. The genome, envelope, and proteins of coronaviruses will influence all steps of the viral life cycle.

The coronavirus is named after its characteristic spike glycoproteins that stick up from the viral surface. These spikes give the virus the appearance of wearing a crown, the latin translation for which is 'corona'. However, unlike the crowns that sit atop of the heads of royalty, the spike proteins of coronaviruses are structurally crucial to the virus's functions—specifically, the attachment and entry steps of the viral life cycle. These spikes consist of an intracellular segment, a segment that passes through the envelope membrane to anchor the protein down, and a segment that reaches outside the viral envelope into extracellular space, called the ectodomain, which is essential to invading host cells. The ectodomain is made up of two subunits. The S1 subunit acts as a key to get inside the cell. The attachment process occurs when the S1 subunit key is inserted in the lock of a specific cell surface receptor. The shape of the key will determine the types of cells it can unlock. Next, the S2 subunit opens the door for the virus to enter the cell by fusing the viral and cell membranes together, allowing the viral genetic material to be injected. For the S2 subunit to function properly, the protein must be cleaved (a bond in the protein must be broken), a detail that will become important in comparing infectivity of coronaviruses (Li).

The replication step of the coronavirus life cycle begins as soon as the genetic material enters the cell. While coronaviruses have the largest genome of all RNA viruses, only parts of their genome can be expressed as proteins. These translatable areas of the genome are known as open reading frames (ORF). Some of these readable windows are translated to make proteins that can replicate the genome. But to form new virions, the coronavirus must do more than replicate its genome. It also needs to build the structural and accessory proteins that make up the body of the virus. To construct these proteins, coronaviruses employ a unique transcription and translation mechanism in which it creates smaller strips of mRNA that each represent an ORF, known as subgenomic mRNA (Wilde et al.). To better describe this process, imagine you were reading a newspaper or magazine. However, the previous owner scratched out some of the sentences in permanent marker, creating gaps in the paragraphs you cannot read. Next, imagine writing down each of the still readable areas on a separate sheet of paper. Each time you come across a scribbled out section, you grab a new sheet of paper and continue copying down the text from there. The readable areas of the original newspaper represent the open reading frames of the genome, while each of the pieces of paper you copied down is the subgenomic mRNA. This analogy is inadequate to describe the entirety and complexity of the process, but the basic idea is that transcription leads to the eventual creation of subgenomic mRNA, which each represents an ORF. The subgenomic mRNA is then translated into the structural and accessory proteins required for viral function.

Assembly of the newly synthesized viral proteins and the copied genome occurs in the endoplasmic reticulum-Golgi intermediate compartment (ERGIC), an organelle that functions as a transport complex. The ERGIC shuttles cellular cargo from the endoplasmic reticulum where proteins are constructed to the Golgi complex where proteins are modified and released (Appenzeller-Herzog and Hauri). While we have briefly described the function of the spike protein, there are three other main structural proteins in coronaviruses. These are the nucleocapsid protein, the membrane protein, and the envelope protein. Membrane, envelope and spike proteins are components required in the virion envelope and are insulated by the endoplasmic reticulum before moving into the ERGIC. Meanwhile, nucleocapsid proteins form the helical capsids of coronaviruses by directly joining the replicated genome in the ERGIC. The interaction between the various structural proteins allows for the assembly of these proteins and genomes into virions that can be released. Release occurs through the formation of tiny vesicles that exit the cell via exocytosis (Astuti and Ysrafil).

Impact of Coronavirus Characteristics on Infectivity

In epidemiology, infectivity is defined as a pathogen's capacity to spread amongst hosts. High infectivity is a key indicator of the epidemic and pandemic potential of a virus. A commonly used metric for the spread of a virus is basic reproductive number (R0), which is defined as the average number of secondary transmissions from one infected person. If the R0 is greater than 1, the epidemic is growing. If the R0 is less than 1, the number of infections is decreasing. It is difficult to obtain an accurate R0 at the start of an epidemic, with COVID-19's estimated R0 ranging from 1.4 to 4.0 (Zimmer). Nevertheless, it is generally accepted that COVID-19 is more infectious than severe acute respiratory syndrome (SARS) and Middle Eastern respiratory syndrome (MERS), two other coronaviruses that have spread to humans. There are many aspects that contribute to infectivity such as the effectiveness of the viral attachment, the virus's entry mechanism, and the timing of viral shedding. These factors all contribute to how SARS-CoV-2 has infected millions of people worldwide, causing one of the largest scale public health crises in recent years.

To begin, we might ask the question: why is COVID-19 spreading so fast compared to SARS? In 2003, SARS had a fatality rate of 14-15% and an R0 of 2.0-3.0; for every one person infected, it would spread to an additional two or three more individuals. SARS is deadlier than COVID-19, yet the novel SARS-CoV-2 is more transmissible (Petersen et al.). While there are many social, political, and economic factors to consider, one of the biological reasons for COVID-19's rapid spread lies in the binding affinity of the virus. Although severe acute respiratory syndrome-related coronavirus (SARS-CoV) and SARS-CoV-2 share very similar structures and pathogenicity, there are slight differences in the spike proteins that can account for differences in infection rates. During the attachment step of the viral life cycle, SARS-CoV and SARS-CoV-2 spike proteins bind to a receptor on human cells known as angiotensin-converting enzyme 2, or ACE2. This is an enzyme highly expressed in the lower respiratory tract, myocardium, bladder, and kidney. Experiments have shown that SARS-CoV-2 spike protein binds to ACE2 with a higher affinity than SARS-CoV spike protein (Mallapaty). In essence, SARS-CoV-2 spike protein "sits" better on human ACE2 than SARS-CoV spike protein and this can increase its ability to attach and enter cells, enhancing COVID-19 infectivity.

Another biological explanation lies within the cleavage of spike proteins that allow for entry into human cells. The SARS-CoV-2 spike protein contains a site that is activated by a host cell enzyme called furin, while SARS-CoV spike proteins do not have this furin cleavage site. Furin is significant because it is found in many human tissues including the lungs, liver, and small intestine. A virus that can take advantage of such an enzyme could gain the potential to be activated by various cell types and attack multiple organ systems. The presence of a furin cleavage site in SARS-CoV-2 could affect virus stability and transmission, although more research is needed to understand how removing the site could modify viral function (Mallapaty).

The timing and length of viral shedding also plays a role in the fast spread of SARS-CoV-2. Viral shedding is the release of a virus from a host into the environment to potentially infect other members of the population. At the start of symptom onset, SARS-CoV-2 is at its peak viral load (the total number of virus particles carried by an infected host). In contrast, SARS-CoV has a viral load that peaks 6-10 days after symptom onset. This allows an extra week to identify and isolate SARS cases, which is highly beneficial in controlling its spread. Viral shedding may also occur for a prolonged period in COVID-19 patients. One study found high viral loads of SARS-CoV-2 in patients up to three weeks after symptom onset. Viral loads that peak sooner after symptom onset, and prolonged viral shedding are factors that contribute to the high transmission of SARS-CoV-2 (Petersen et al.).

Structural characteristics of SARS-CoV-2 directly impact completion of its viral life cycle, with variations in spike protein affinity, cleavage, and viral shedding affecting the steps of attachment, entry, and release. Uncovering the reasons for high transmission of COVID-19 is an arduous process. The biology of SARS-CoV-2 can be complex and results of experiments can be difficult to interpret. Regardless, scientists around the world are racing to understand the virus, testing antiviral drugs and vaccines at record speeds to protect humanity against rapid COVID-19 spread. Understanding of the factors that contribute to infectivity allows policy makers to do their best in mitigating transmission. Viral physiology may only be a part of the puzzle, but undeniably a crucial one. Thus, the race of scientific researchers against time continues.

CHAPTER 8

SYSTEMIC RESPONSE TO COVID-19: INTERVENTION PORTFOLIOS

BY TINA WU

As the largest pandemic thus far of the 21st century, COVID-19 has had massive implications on public health systems worldwide. The pandemic has not only caused widespread fear and panic, but also fundamentally changed the way we work, maintain relationships, and live our lives. Countless individuals are emotionally exhausted and itching to return to the norm. A decrease in coronavirus cases starting May of 2020 has left many Canadians hopeful. Several questions linger in the back of our minds: when will the pandemic be over? And what can our governments and communities do to move the process along?

The End of the Pandemic

The World Health Organization (WHO) will likely announce the end of the pandemic once the coronavirus infection rates drop significantly worldwide and the virus becomes reasonably self-contained. The span of time it takes for the pandemic to cease depends on the actions of governments globally. In a TED-Ed video with over 2 million views, three main strategies that governments and communities can employ to contain and end a pandemic are described: race through it, delay and vaccinate, or coordinate and crush (Rosenthal).

At first glance, racing through a pandemic to quickly reach an end seems favourable. This strategy is engaged when governments and communities allow the population to be exposed as quickly as possible and do not attempt to mitigate the spread of the virus. Unfortunately, this method comes at the cost of many lives. If scientists are not given adequate time to study a novel virus, doctors may not be able to treat it. The uncontrolled spread of the virus leads to hospitals filling up to capacity almost immediately, potentially leading to millions of deaths and the collapse of the healthcare system entirely. Survivors will develop an immunity to the virus. Once the majority of people have been infected, herd immunity prevents the virus from infecting new hosts and the pandemic ends as quickly as it began. However, this method will not work if people can get reinfected (Rosenthal).

There are other methods to ending a pandemic that do not have such a high cost of life. Delaying and vaccinating involves slowing down the spread of the virus to provide time for researchers to develop a vaccine and ensure that hospitals can manage the number of infected patients. The measures taken might include physical distancing, widespread testing, and quarantining the infected. Yet even with these measures in place, outbreaks can still occur causing up to hundreds of thousands of deaths. Additionally, if a city appears to have outbreaks under control and resume normal business, there may be a resurgence of cases (Rosenthal).

Finally, coordinating and crushing the outbreak involves treating the world as an interconnected system and simultaneously starving out the virus everywhere. The key is for all nations to synchronize responses, in a combination of quarantine, travel restrictions, and physical distancing. This could end the pandemic in just a few months with a low loss of life. While ideal in theory, it can be difficult to get all world powers to act simultaneously. When one country is peaking, another may only be starting to have an emergence of cases. It would be incredibly difficult to convince a nation to go into complete lockdown, considering the economic and social consequences, if only a small fraction of the population is presenting with the condition. If the virus is not completely eradicated during this time, the risk of resurgence to pandemic levels continues. If animals can spread the virus, efforts of coordinating and crushing the virus may also be undermined (Rosenthal).

Each of these responses have their benefits and drawbacks. Racing through the pandemic is a quick fix but at a high cost of life. Delay and vaccinate involves a smaller number of fatalities, but fatalities nonetheless. Coordinate and crush is effective but may not be feasible if world leaders cannot act simultaneously (Rosenthal). Different countries have deployed different tactics to reach the end of the pandemic. The specifics of intervention portfolios will vary from country to country.

Systemic Response in China

China's intervention on COVID-19 can be described as the coordinate and crush tactic, although epidemiologists identify a glaring flaw in the response—it occurred too late (Cyranoski). When Dr. Li Wenliang first noticed a SARS-like virus in his patients near the end of December in 2019, he alerted several of his colleagues through a Chinese social media platform, Weibo, advising the other doctors to wear protective clothing to avoid infection. Four days later, he was paid a visit by the Public Security Bureau and signed a paper that stated he understood the consequences of continuing to spread fake rumours and disturbing the social order. It was more than two weeks later, on January 20, that China recognized the situation as urgent and declared the outbreak an emergency (Hegarty).

On January 23, Wuhan, the centre of the epidemic, was put on lockdown. 15 other cities in the Hubei province followed and nearly 60 million people were affected. Roads were blocked, and flights and trains were suspended (Cyranoski). Citizens were encouraged to stay at home and only leave when necessary to get food or medication. A New York Times article describes that housing complexes of certain cities issued the "equivalent of paper hall passes to regulate how often residents leave their homes" (Zhong and Mozur). WHO applauded China for its unprecedented public health response that reversed the escalating cases in a report in February. Before public health measures were introduced at the start of the pandemic, scientists estimated that each infected person would pass the virus on to at least two other people. After the strict quarantine was imposed, this number dropped to 1.05 (Cyranoski). On April 8, 2020, the Wuhan lockdown officially ended (Zhong and Wang). Early modelling suggested that without any efforts at containment, the virus would spread to at least 40% of China's population—500 million people (Cyranoski). As of August 9, 2020, just under 85,000 cases have been reported ("China"). Some scientists believe that there exist a lot of unreported cases, but even if the true number of cases was 20 or 30 times the reported amount, this would still indicate that China's tactics were successful in combating COVID-19 spread (Cyranoski).

Canadian Interventions on COVID-19 Spread

Canada uses the delay and vaccinate model on intervention for COVID-19 by implementing travel restriction policies, widespread testing, the shutdown of non-essential businesses, the closure of schools and universities, education on hygiene practices, contact-tracing, and the quarantine of infected individuals.

The Canadian Health Network outlines a timeline of government actions regarding COVID-19. In the early stages of the pandemic, Canada introduced screening at airports to identify individuals flying in from China that have flu-like symptoms. When the first Canadian case of COVID-19 appeared in a man from Toronto in late January of 2020, public health officials attempted to trace those who might have been in contact with the patient during his flight. In the end, the man and his wife were the only two people infected. The husband was isolated in a negative pressure room at Sunnybrook Hospital, while the wife self-isolated at home, both making full recoveries (Bronca).

In February, travel-related precautions remained fixed on individuals arriving from China. The nation's top public health official, Dr. Theresa Tam, instructed travellers from the Hubei province, where the outbreak began, to limit contact with others for 14 days upon arrival. Travellers were educated on the actions to take if they fell ill with the disease, such as self-isolating and informing public health, but Canada held out against travel bans. At the same time, the wider population was educated on the importance of proper hygiene practices including hand-washing, and avoiding contact with one's eyes, nose, and mouth (Bronca).

By the end of March, Canada's response to travel had drastically shifted. Two days after the WHO declared COVID-19 a pandemic, Health Canada advised to limit non-essential travel. Subsequently, all travelers entering Canada were asked to self-isolate for 14 days. On March 18, Canada went a step further, putting a ban on all foreign nationals entering Canada. Passengers showing symptoms of COVID-19 were prohibited from boarding flights into the country, and penalties were given to travellers who did not self-isolate for the mandatory 14 days. Throughout March, testing was focused on travel-related cases. But as the number of people presenting with symptoms flooded the system, massive designated assessment centres were opened. Provinces ramped up testing to try to meet the demands, with many cities launching self-assessment tools to aid in the process. By the end of March, Canada had processed over 100,000 tests (Bronca). As of August 6, Canada has processed 4.3 million COVID-19 tests (Elflein).

March is also the time when provinces began declaring states of emergency. The NBA and other sports leagues postponed their seasons indefinitely, schools and universities closed in favour of online learning, and the provinces of Quebec and Ontario ordered the shut-down of non-essential businesses (Stone et al.). Social distancing became common practice. Yet, it was not until late May that Canada's chief public health officer began recommending non-medical face masks and coverings to prevent COVID-19 spread (Gilmore). Before this time, it remained relatively unclear whether the masks would be effective in combating spread, but as new information surfaced about asymptomatic transmission, it became commonplace to wear masks in public areas. With all these measures in place, the number of coronavirus cases began to plateau in May. Provincial governments cautiously reopen the economy, with many provinces doing so in stages ("Reopening Ontario"). As stores resume activity, many provide social distancing guidelines, hand out face masks, and dispense hand sanitizer at the store front.

Newer tactics to combat COVID-19 include the launch of a free exposure notification app. The COVID Alert app is a method of contact tracing. It works by using Bluetooth to connect with other phones with the app running in the background. If a user tests positive for COVID-19, they receive a one-time code that is sent through the network to anyone who has been within two meters of the user for more than 15 minutes within the past 14 days. Unfortunately, a Toronto Star article written August 5th reports that less than 4% of the country's population has downloaded the COVID Alert app despite having better privacy protection than Facebook. Experts have previously stated that at least 60% of the population needs to have the app installed for it to be effective in the midst of a pandemic (Saba).

In summary, Canada has responded to the pandemic by implementing travel restrictions, testing when possible, shutting down non-essential businesses, educating the public on hygiene practices, closing schools and universities in favour of remote learning, quarantining or instructing the self-isolation of infected individuals, and launching a new app that is yet to be determined effective. All these methods are meant to "flatten the curve" of novel coronavirus cases to an infection rate that is manageable by hospitals and healthcare practitioners, while waiting for research on antiviral treatments, therapeutics, and vaccines to emerge.

Systemic Response in the United States of America

The systemic response to the novel coronavirus in the United States has been heavily criticized. Despite the heavy funding for vaccine research, many believe that the Trump administration has not done enough to delay the virus's spread. The nation might be racing towards the end of the pandemic, rather than delaying and vaccinating. As of August 9, 2020, the US sits at 5.05 million confirmed cases, which represents a quarter of the COVID-19 cases worldwide despite having only 4% of the world's population .

The Atlantic writes a scathing article on the government's response to the pandemic: "The breadth and magnitude of its [America's] errors are difficult, in the moment, to truly fathom" (Yong). The government's early response to the novel coronavirus involved closing borders. However, travel bans can only slow the spread of the virus, not prevent it. The article rides on this idea and claims that the crucial early weeks could have been used to mass-produce tests, or manufacture protective equipment and ventilators. Instead, the focus was solely on border control. The CDC developed and distributed tests in late January, but these tests had to be recalled due to a faulty chemical component. Tests created by private laboratories were strangled by FDA bureaucracy. In April, four out of five frontline nurses said they did not have enough protective equipment. Where the government was ineffective, the people acted. Businesses sent their employees home and people began social distancing even before the national state of emergency was announced on

March 13. A June survey found that 60-75% of Americans were still practicing social distancing. This is hopeful as Beth Redbird, a sociologist at Northwestern University who led the survey says "in public-opinion polling in the U.S., high-60s agreement on anything is an amazing accomplishment" (Yong).

On the vaccine research side, the US has made progress. A regularly updated vaccine tracker on the New York Times provides insight into United States government spending. Moderna, in partnership with National Institutes of Health, created the first vaccine candidate to be used in human trials in March. It launched its phase III trial on July 27th, enrolling 30,000 healthy people at 89 sites in the United States. The company was granted government funding of nearly one billion dollars for its research. The Trump administration has also awarded a 1.9 billion dollar contract to German company BioNTech, the New York based Pfizer, and Chinese drug maker Fosun Pharma for 100 million doses of the vaccine by December. They are currently within a combined phase II/III trial. A final notable mention is Novavax, which launched its coronavirus vaccine trials in May. The US government has awarded 1.6 billion dollars (Corum et al.).

Conclusion

Systemic response to COVID-19 varies from country to country as different policies and plans are enacted to bring us closer to the end of the pandemic. Some countries have opted to race through it, others delaying the spread to allow the research to catch up, and others still have set strict regulations to squash the disease. With the collective efforts of governments and communities, the pandemic will come to a close. If we learn from our successes and failures, the next time a highly contagious virus shakes the globe, we will be prepared.

CHAPTER 9

TRAVEL RESTRICTIONS AND THEIR IMPACT ON EXISTING BIOSOCIAL SYSTEMS

BY GRACE LETHBRIDGE

COVID-19 has spread rapidly around the globe since it was first identified in Wuhan, China in December 2019. Wuhan, a city of 11 million people, first diagnosed four cases with what they believed was viral pneumonia, but one not responding to typical treatments. By the end of the month, dozens more fell ill. It is now understood that the number of people infected was much larger than what they had originally thought. It is now conservatively estimated that 1000 people were infected by the virus at that point (Wu et al.). The World Health Organization (WHO) officials were notified by Chinese officials of the disease on December 31, 2019. Although reassuring the public it was controllable, it is believed the reproduction number was between one and three, indicating that each infected person was spreading the disease to another one to three people. The timing of this outbreak is significant because it occurred during the Lunar New Year when many people were travelling back to their hometowns to visit their families. According to cell phone traffic, seven million people travelled out of Wuhan before travel restrictions out of the city were enforced on January 21, 2020 (Wu et al.). By then, COVID-19 had already spread to major cities in China, as well as to cities around the world, with the first case outside of China reported in Bangkok in mid-January. By the time other countries started banning flights from Wuhan, it was already too late.

COVID-19 is a highly infectious virus, where infected individuals are contagious while asymptomatic or pre-symptomatic. Asymptomatic patients are believed to account for 40% of the total infected population (CDC). This is a serious public health challenge because many carriers of the virus continue to spread the disease unknowingly. This type of virus, coupled with global travel, was the perfect setting for the conception of a pandemic. Although many individuals have criticized the efficacy and lack of containment strategies enacted by Chinese officials at the beginning of the spread, it is important to recognize that it would have been nearly impossible to prevent the spread beyond Wuhan without having a solidified universal plan for managing infectious diseases. When the outbreak first occurred in Wuhan, there still remained much unknown about the virus and its propagation; therefore, the long-term global impact of the virus was unforeseen. The interconnection created through global travel further exacerbated the spread of the virus, ultimately leading to the pandemic. Travel has since been restricted in varying ways by different countries and jurisdictions, further complicating the impact on individuals and communities.

As borders were closing, people across the globe were scrambling to determine whether they could or should shelter in place or return to their country of citizenship. On the surface this seemed somewhat straightforward, but for many groups of people it was not. This chapter will explore the groups of people who were impacted by these travel restrictions.

International Students

International students are typically in Canada on a student visa, with no family or social network beyond their schools. As many universities suspended all in person activities in March, international students were struggling with the decision to either return to their home countries or to find permanent housing in the country they were studying in. Many universities in Canada helped these students travel home, find alternate housing, or provided them accommodation in a student residence. However, some were unable to enter their home countries due to travel restrictions, leading to them unexpectedly separated from their families, unable to work, possibly incurring debt, and ineligible for government assistance (e.g. Canada Emergency Response Benefit, Canada Emergency Student Benefit). Other students chose to stay in Canada for a variety of reasons—higher rate of infection in their home country, staying to work in Canada, or concerns with being unable to re-enter Canada for the coming school year. The students who stayed in Canada, regardless of why, may now be in a less complicated position than those who left.

As universities begin to reopen in September, whether that be online, in person, or a combination of both, international students are once again impacted. Due to travel restrictions and varying public health restrictions across Canada's provinces, international students may be unable to return to Canada unless they can prove they must return in order to continue their education in person (i.e. schools are requiring in-person classes and/or labs, unreliable internet service in their home countries, etc.). International students who do not meet these requirements will be required to complete their coursework online or temporarily suspend their studies. International students are an integral part of Canadian universities and colleges, so it is crucial that the government assesses the impact this may have on prospective international students as well as the international tuition rates many universities rely on as part of their funding.

Family Separation

It is not just international students who have faced family separation due to COVID-19 travel restrictions. As the world has become more of a global community, families themselves have become global. Although Canada allows for reunification of families across borders, it may not be straightforward for couples living in a common-law relationship. Unless the non-citizen partner has permanent residency, they must prove their common-law status to be allowed entry into Canada. Canada has taken a more liberal approach to family reunification than many other countries. For example, in Israel, the Orthodox Chief Rabbinate controls all legal Jewish marriage, and under Orthodox Jewish law, interfaith and same-sex marriages are forbidden. To overcome this, secular, LGBTQ+, and interfaith couples have opted to marry elsewhere and then later file the paperwork to have their marriage certified in Israel. Others have chosen not to get married. For couples who never left Israel to get married, or for those who did, and had yet to file the paperwork to have their marriages certified, these families have been further disadvantaged by COVID-19. Partners and children outside of the country, without Israeli citizenship, are indefinitely separated from their partner and parent in Israel ("For Some Israelis, Covid-19 Has Meant Separation from Their Families").This further isolates and segregates individuals who are part of these minority groups, exemplifying the complexities surrounding the cascading effects of COVID-19.

Travel to Avoid COVID-19

Although travel has been a main contributor to the pandemic and has created many barriers to education and reunion of families; for some, travel has been a way to avoid COVID-19. Many people with financial privilege have been able to use their wealth to travel outside of COVID-19 hotspots. During the outbreak in New York City, wealthier residents were able to travel to second homes in other regions of the country. With fewer restrictions on this type of travel, these travellers could have been putting others at higher risk. Beyond the higher mortality rates among the elderly and those with comorbidities, infection rates are much higher among those in lower socio-economic classes. In particular, black, Indigenous, and people of colour (BIPOC), often working in low wage, essential jobs, and living in tighter quarters, are the people put at further risk by wealthier individuals escaping COVID-19 hotspots.

A similar phenomenon exists in Canada. Early restrictions were enforced to prevent people from traveling to their cottages at the beginning of the summer. Cottage regions do not have extensive medical infrastructures to withstand an outbreak. There were concerns that an outbreak in one of these regions would overwhelm health services and leave the permanent residents vulnerable. Cottage ownership is often the domain of the upper middle class and the wealthy, while many permanent residents are those working in service jobs to support the cottage industry. Though many Canadians were supportive of this approach; for some cottage owners, they felt it to be a violation of their rights as property owners and to travel within their own country or province.

Beyond the wealthy being able to travel to second homes and cottages within their own countries, wealthy individuals with dual citizenship can and do travel to the country of citizenship they feel is the safest for them.

This ability to lessen the risk of contracting COVID -19 by traveling elsewhere is a luxury afforded to the few who can afford it. Income disparities continue to fuel health inequities; therefore, it is imperative that the government implements policies to address the social determinants of health.

Tourism Industry

Travel is ubiquitous. People travel for business, for pleasure, to connect with family, and to transport essential goods and services within and beyond their own countries. During the pandemic, a lot of emphasis has been placed on ensuring the necessary travel of essential goods and services and on restricting travel for non-essential business and pleasure. This prudent approach is accompanied by consequences for the millions of people across the world who work directly or indirectly in the tourism industry. These workers are mostly lower paid service workers, such as wait staff, flight attendants, hotel cleaning staff, tour guides, etc. who are unable to sustain unemployment, especially without government support. It is also still unclear what the long-term implications will be to non-essential travel. Reopening the industry will be more challenging than closing it down was. Although the primary focus has been to develop vaccines and treatments which would aid in reviving the tourism industry, the long-term impact of the pandemic on consumer behaviour still remains unknown.

Many large corporations within the tourism industry are emphasizing the importance for consumer safety through stringent hygiene measures; however, it is unclear whether the public trusts these corporations with their safety — especially businesses such as cruise lines which were key contributors to the spread of the disease. Additionally, given the global economic impact of the pandemic, tourism is seen as a luxury, so will people be willing to spend money on a luxury they may be unable to afford or simply do not trust? The impact of this mentality shift on the tourism industry and the millions employed within it remains to be seen (OECD).

COVID-19 has magnified the social inequities that led to disproportionate negative health outcomes among disadvantaged groups. Travel, as both a necessity and a privilege, has contributed to the spread of COVID-19. It has also allowed those of privilege to avoid the disease while potentially putting poorer communities at risk. Global travel, and its role in the spreading of a contagion, requires a global plan to ensure societies are better prepared to respond to the next outbreak.

CHAPTER 10
DATA CHALLENGES AND SPACE-TIME PREDICTIVE MODELS

BY GRACE LETHBRIDGE

COVID-19 Data Challenges and Space-Time Models

Coronavirus disease 2019 (COVID-19), caused by the severe acute respiratory syndrome coronavirus 2 (SARS-CoV-2), was declared a global pandemic on March 11, 2020 by the World Health Organization (WHO). The virus was first described in December 2019 in Wuhan, China, which quickly evolved into a global pandemic within three months. The unpredictable and rapid spread of COVID-19 led to major global changes. To respond to the ongoing pandemic, many countries implemented measures, such as self-isolation, physical distancing, and mandatory public use of face masks in order to prevent the spread of the disease. The objective of these measures was to flatten the curve, which consequently mitigates the burden on healthcare systems.

The COVID-19 pandemic requires the collaboration of researchers, healthcare workers, epidemiologists, and policy makers. The rate at which COVID-19 spreads and to where it spreads are important variables to investigate because they influence public health decisions by governments to address the multivariate effects of the pandemic. Policy makers rely on data collection and space-time models to inform best containment strategies, such as patient quarantine, contact tracing, border controls, and community education and regulations.

The data supplied to underline health policy decisions must be accurate, standardized, and abundant. This data should be collected through reliable COVID-19 testing and tracing on regional, provincial, national, and global scales; however, amidst a pandemic, this is difficult to achieve. There are many methodological challenges involving the production, collection, analysis, reporting, and publication of COVID-19 data with shortened timelines (Wolkewitz and Puljak). The shortened timelines are further challenged by the need to gather data across multiple jurisdictions, in real time, and to adjust the models accordingly. These data challenges impact the quality of COVID-19 data collected and subsequently the models produced to predict the future spread of the disease.

The following challenges impact the quality of clinical and epidemiological studies regarding COVID-19: standardized clinical data acquisition and storage and common statistical errors in clinical studies.

The lack of standardisation in clinical data acquisition and storage may impact the quality of future clinical cohort studies. Cohort studies rely on highly structured and refined data wrangling techniques in order to draw objective conclusions. Once the data has been merged and cleaned, artificial intelligence, machine learning algorithms, and deep learning algorithms can be applied to learn and predict future outcomes. Another issue that arises is there is not one singular universally accepted algorithm that can be used to predict future COVID-19 cases in various countries. A study, performed by the University of Patras, suggested algorithm performance may vary from country to country due to the following factors: country specific climatic and geographical characteristics, various population densities among countries, disparities in testing and measuring procedures among different countries, and diversity in regards to the duration and severity of policies on quarantine and social distancing across different countries (Papastefanopoulos et al.). Another result from the lack of standardisation in clinical data acquisition and storage is the unclear distinction between active and closed COVID-19 cases. Ideally, active cases refer to patients who are currently hospitalized and closed cases refer to patients who have died or have been discharged. However, the lack of standardisation impacts the collection and specificity of these quantities. Another barrier to supplying rapidly available clinical data is the timeliness of administrative protocols, such as patient consent forms, ethics statements, and data security measures. Although these protocols are very important, they need to be streamlined in order to meet the data demands of an emerging pandemic.

Common statistical errors in clinical studies include length-time bias and competing risk bias. Length-time bias is a type of selection bias in which there is an overestimation of survival duration due to an excess of asymptomatic cases with slow progression detected through screening, while cases with fast progression are detected after symptoms have begun. This bias makes it seem as though the screening caused the different outcomes,

80

rather than the disease progression – an important consideration when looking at COVID-19 testing, incubation periods, proportion of cases that are asymptomatic, and duration of infectivity. Competing risks are events which prevent the occurrence or alter the risk of the primary outcome of interest (Lesko and Lau). Calculated cumulative incidence and predicted risk values may be biased upwards if competing risks are ignored (Abdel-Qadir Husam et al.). In the context of COVID-19, competing risk bias may impact reported values for mortality and cure rates. Therefore, it is imperative that statistical analyses of COVID-19 data account for these biases.

COVID-19 space-time predictive models provide relevant information on disease transmission. Predictive models provide estimates on the rates of infection, hospitalization, recovery, and mortality with respect to time. These forecasts, combined with spatial analytics, can detect potential clusters of COVID-19, allowing time for informed prevention and control decisions and actions. The University of Pennsylvania's Predictive Healthcare Team created the CHIME model (COVID-19 Hospital Impact Model for Epidemics), adapted from the susceptible, infected, and recovered (SIR) mathematical model, to estimate the quantity of people to be admitted to the hospital, as well as the number of ICU beds and ventilators needed. Models like CHIME can then be integrated with geographic information systems (GIS) which facilitate the visualization of geographical pandemic information, spatial tracking of confirmed cases, prediction of regional transmission, and management of the supply and demand of resources (Zhou et al.). GIS users can choose specific start dates and the average number of days of infection to project the future geographical spread of the disease (GISP). The models used with GIS have various parameters that can be adjusted to examine how the forecast changes in response to different scenarios, such as physical distancing and quarantine measures. The visualization of health policies' effects across space and time continue to inform important policy decisions regarding COVID-19 prevention and control.

COVID-19 and Biosocial System Dynamics

The Syndemics model of health focuses on how biosocial systems influence the co-occurrence of diseases.This model examines why specific diseases tend to cluster in individuals and populations. It examines the interactions of diseases through biological pathways, as well as how social environments, such as social inequity and injustice, contribute to disease clustering and interaction (Lancet).

The spread of COVID-19 in Canada was not random. The spread of COVID-19 relies on the synergistic interaction of biological, political, social, and economic systems. People with the following conditions are at increased risk of severe illness from COVID-19: cancer, chronic kidney disease, obesity, diabetes, respiratory conditions, and heart conditions (CDC). Therefore, in the context of COVID-19, it is important to consider the public policies and social environments that interact to overlap these disease clusters. The most severe and concentrated cases of COVID-19 occurred in long-term care homes and lower income urban neighbourhoods, especially racialized communities. Low income racialized communities experienced disproportionate high rates of infection (CDC). The unavailability of affordable housing has caused low income individuals to aggregate in overcrowded apartments. Many lower income individuals are also deemed "essential workers" due to their work as clerks in grocery stores or as personal support workers. The combination of living in overcrowded housing with increased exposure to infection, as essential workers, significantly increases the risk of infection in these populations. These social problems generate concentrated clusters of diseases in vulnerable populations. Going forward, Canadian social infrastructures must be transformed in order to address the health inequities experienced by vulnerable groups.

Biosocial System Dynamics' Impact on Modelling

As epidemiologists and researchers continue to develop mathematical models to predict, monitor, and visualize the impact of COVID-19 on societies, the demand for models that examine biosocial influences is ever growing. The processes that determine COVID-19's dynamics are often classified into two components: biological and societal social management. Biological processes determine the basic reproduction number (i.e. potential number of people to be infected by one person), probability of recovery, and visibility or severity of symptoms. Societies can manage the pandemic by limiting the contacts of those infected through self-isolation or by enacting social distancing measures and mandatory mask use in public spaces. Mathematical models rely on foundational assumptions in order to generate estimates. Different models use different assumptions. Common assumptions in COVID-19 models are the following:

- *Biological functions are the same and constant in time*
- *The virus does not mutate*
- *Asymptomatic carriers of the virus are no longer contagious after 16 days*

Common factors that many COVID-19 models ignore are the following:

- *Potential immunity in recovered patients*
- *Population structures (i.e. sex and age structure, socioeconomic classes, infrastructure, etc.)*

These assumptions and exclusions may generate flawed mathematical models; the precise models that inform the decisions of policy makers. Social systems are variable, and mathematical models require further sophistication to account for biosocial influences. Sex and age structure, socioeconomic classes, and infrastructure are important determinants in rate of infection. Many COVID-19 clinical and epidemiological reports include the increased mortality rates in people with pre-existing conditions; however, these reports frame the pre-existing condition as justification for mortality, rather than exposing it as a synergistic factor of the disease trajectory itself (Herrick). Older people or people with pre-existing conditions are classified as "vulnerable" individuals. This language erases the politics of those vulnerabilities and the origin of the pre-existing conditions themselves. When the models do not account for the varying socioeconomic, environmental, and cultural disparities within societies, it becomes increasingly difficult to analyse and interpret the data generated. Further questions arise: Are two individuals, both with pre-existing hypertension, at the same risk of COVID-19 related mortality, independent of their socioeconomic class? More sophisticated COVID-19 mathematical models that address the social structures influencing the genesis of pre-existing conditions are needed to better interpret forecasts.

Data Informed Decisions

COVID-19 has impacted the world in numerous ways. It has caused over 600,000 deaths, devastated economies, collapsed healthcare systems, and weakened global connectivity. The estimates generated from predictive models have informed many of the global policy changes regarding isolation, lockdowns, social distancing, masks, border control, and allocation of resources. When predictive models are used to inform these changes, it is imperative that these models are mathematically sophisticated, accurate, and interpretable. A notable example of the dangerous effects of using problematic models was when Neil Ferguson, from the Imperial College London, released his epidemiological model on March 16, predicting that tens of millions of people would die from the COVID-19 pandemic. Due to his affiliation with the Imperial College London, his model was considered a credible source of information. Shortly after the release of Ferguson's model, the US-Canadian border closed to non-essential travel and the Prime Minister, Justin Trudeau, announced an $82-billion aid bill to address the loss of jobs expected from the pandemic. Though the release of Ferguson's model cannot be causatively linked to the changes in border control and aid, the timing of the events is suspicious. The changes themselves are not what is in question, rather the lack of high-quality evidence to inform those decisions and large-scale changes. Months later, serious flaws in Ferguson's model were uncovered. The lack of data addressing the biosocial systems involved in the spread of COVID-19 is dangerous, as it can underestimate the impact on vulnerable communities. The dangerous effects of poor projections are endless, which demonstrates the need for critical appraisal of clinical and epidemiological literature. Policy makers rely on scientific models and forecasts; therefore, it is crucial that scientific models and computer code are properly scrutinized and audited.

CHAPTER 11
SOCIAL MEDIA

BY LILY LIU

In the era of the COVID-19 pandemic, the different types of social media have been a key tool for communication and collaboration under extraordinary circumstances. Sites and apps such as Zoom and Google Meet have seen a spike in new users attempting to continue conducting their usual professional affairs, and other popular social media outlets - such as Twitter, Facebook, and Instagram - have also had their activity boosted by the isolation period that comes with practicing social distancing measures. Social media has been used as a platform for remote communication, an instrument to spread information about the COVID-19 virus, and a form of entertainment during the difficult and monotonous routine of quarantine. While the use of social media comes with its benefits and consequences, it is undeniable that its presence and prominence have been one of the most significant factors of the COVID-19 pandemic in 2020.

The existence of social media is, first and foremost, intended as a means of communication in a modern-day age where more traditional forms of social interaction have been replaced and upgraded with new technology. Its common presence and convenient properties have given rise to a revolution in information. As is strongly recommended in the COVID-19 pandemic, social distancing measures have been implemented in nearly every affected nation in the globe, and the workings of international society have halted as a result of this movement - citizens are no longer travelling outside their home, attending educational institutes, or going to their workplace for their careers. This has fundamentally changed the way that companies, corporations, institutions, and other businesses interact with their employees. Face-to-face communication between individuals is no longer possible due to these measures, which require people to isolate themselves from the majority of outside society and only leave their homes when absolutely necessary, and many events and in-person activities have been cancelled or outlawed in order to reduce any risk of the virus spreading.

The main reason that social media is so crucial during the COVID-19 pandemic is its ability to provide a remote means of communication to individuals who are complying with social distancing measures and staying at their homes. In lieu of in-person interaction, platforms that offer messaging services such as Telegram, Facebook, and Gmail allow people to make contact and touch base with the acquaintances they aren't able to meet. These platforms' simple interfaces and convenient features make them easy to use and advantageous for all forms of communication. Furthermore, in place of the in-person meetings and classes that are usually a significant part of the operation of businesses and educational institutions, social media applications that allow video and voice call - such as Zoom, Google Meet, Whatsapp, and Skype - can grant its users the capability to collaborate with fellow employees or students regardless of their location. In the era of quarantine and social distancing, these forms of social media are arguably the most crucial in permitting society to function as closely as possible to how it did previously, before the beginning of the COVID-19 pandemic and its eventual spread around the globe.

The increased use of social media platforms, however, has also led to a rise in the spread of information in regards to the COVID-19 pandemic. As the primary international issue in the world, a great deal of attention has been focused on the spread of the virus and the details surrounding its properties - for example, the virus' symptoms, how it is transmitted from human to human, and the ways its escalation can be prevented or, in other words, the method of "flattening the curve". This movement has given way to an outbreak in misinformation and inaccurate statistics and data concerning COVID-19 as the various news and figures around the pandemic see a sharp growth. Tedros Adhanom Ghebreyesus, the director-general of the World Health Organization (WHO), referred to this outbreak as an "infodemic", a mass movement that generates widespread fear and panic around the COVID-19 virus and is amplified by social media.

One example of the misinformation spread through this "infodemic" includes the belief in a conspiracy theory that the virus was developed as a means of biological warfare by China, the country home to the epicentre of the COVID-19 pandemic, although there was little to no data or proof to back up this claim. Social media was a key factor in spreading rumours related to this idea, which led to unnecessary sentiments of fear and hostility among members of the online community. The COVID-19 virus being implicitly associated with China has also been a case where negative ideas are heightened by the rapid spread of information through the Internet. This is a significant downside to the convenient, readily available nature of social media; users posting inaccurate or untrue information are able to have a platform on which to spread their misleading beliefs, which can cause an unwanted culture of anxiety and hysteria among members of the general public who regularly access online communities and websites.

In addition, another case where the "infodemic" factor comes into play is the misinformation surrounding the handling of the pandemic itself. Multiple unconfirmed and questionable pieces of medical advice have become omnipresent as a result of their circulation around social media - for example, the belief that COVID-19 may be cured through unconventional means such as the consumption of substances like garlic soup, bleach, cocaine, cow urine, spices including saffron and turmeric, and hydroxychloroquine, a chemical used in medication for treating malaria. These rumours are spread around the Internet with no substantial statistics to back up their supposed ability to cure the virus. In the case of hydroxychloroquine, the misinformation led to a man in Arizona dying after he ingested a similar chemical named chloroquine phosphate under the belief that it would protect him from being infected with COVID-19. This proves that the "infodemic" is in fact a highly dangerous movement that has the capability to threaten the lives of the individuals who buy into its myths and falsehoods.

In other cases, the spread of misinformation through social media can often compete with accurate facts and statistics in terms of their influence on the public. This may lead to people being affected through believing incorrect information over information about the virus that has been scientifically proven, which, by extension, contributes to the mass fear, panic, and general unrest surrounding the COVID-19 pandemic. One such example is the rise of the anti-mask movement that began after numerous countries enacted laws requiring their citizens to wear medical masks. In essence, masks are intended to prevent the rapid spread of COVID-19 by restricting the flow of respiratory droplets that are formed when an individual coughs or sneezes. However, people who are against the use of these masks - commonly dubbed "anti-maskers" -

refuse to wear them and claim that they cause breathing difficulties and do not perform their purpose as well as medical professionals believe they do.

Social media is a large part of this movement because the spread of these misinformed ideas are accomplished through publishing articles on the Internet, posting in widely-used apps such as Facebook and Twitter, advertising rallies and protests against the use of masks, and uploading content such as videos or infographics on sites such as YouTube. In one such situation, "Unmasking America", a large anti-mask group on Facebook with over 9,000 members, was suspended and removed by the social media site on the grounds of spreading misinformation surrounding the pandemic. Facebook cited violation of its policies in promoting inaccurate news as the reason for its suspension. By eliminating the platform that members of the movement use to broadcast their misleading claims, the risks of a social media induced infodemic are reduced, and the advancement of information that has been scientifically backed and proven will be promoted as a result.

On the other hand, social media has also played a positive role in terms of its benefits during the era of the COVID-19 pandemic. The spread of accurate and helpful information can also be achieved through taking advantage of widely-used applications and websites such as Facebook and Twitter. Because these platforms have an extremely large audience of users, information that supports the science surrounding the virus may also be broadcasted to individuals who require it. Furthermore, in the age of social distancing and quarantine, many people see the use of social media as a method of escape from the difficult reality of the pandemic. Communication efforts are streamlined through the Internet, and entertainment may be found in these platforms in the form of remote chats and video calls between individuals who are unable to meet under the current circumstances.

CHAPTER 12
SOCIAL BEHAVIOURS

BY LILY LIU

COVID-19 is a virus that has affected the globe and led to a multitude of dramatic changes across international society. From upending everyday life to drastically impacting the state of world economics and politics, the pandemic surrounding this novel coronavirus has become a point in history that will remain significant through the ages. The most prominent consequences of the COVID-19 pandemic include those that influence the daily lives and mental state of the people around the world. Due to its extraordinary nature and considerable effects, the lives and behaviours of many have been direly altered - in terms of changes in social interaction, everyday communication, mental health and well-being, and personal response to measures that were taken as a result of the pandemic, the reach of the pandemic has extended beyond the surface of everyday life through its reverberations on the social behaviours of individuals who have been affected.

In terms of the negative influence that the virus has had on social behaviours, phenomena such as fear mongering, panic buying, and the drastic behaviour of those unwilling to comply with social distancing measures are the most significant and have been recorded since the beginning of the pandemic. As discussed in the previous chapter regarding social media, the rapid and omnipresent spread of misinformation around the virus is one of the most problematic parts of the general international response to the COVID-19 pandemic. Social behaviours are an integral part of the use of social media; by publicizing false information on platforms that are used by millions of people every day, we risk affecting the mental states, thoughts, and opinions of individuals who are able to view those pieces of information. The "bandwagon effect" is a widespread consequence of the misinformation movement, which has been dubbed an "infodemic". Because people are easily influenced by the words and actions of those around them, they may subconsciously or knowingly imitate and replicate those movements, which would then go on to allow further circulation of the inaccurate ideas they perpetuate. This is a significant factor in terms of the impact that COVID-19 has had on social behaviours among the population.

The COVID-19 pandemic has also been shown to be a traumatic event that greatly impacts the mental and emotional states of those who it affects.

As a result, a mass movement of fear, panic, and general disorder has risen as a response to the handling of the virus. These may include social behaviours such as fear mongering, which ties into the "infodemic" of spreading false and inaccurate information - dramatic and overexaggerated rumours and theories concerning COVID-19 have continued to be spread throughout the pandemic, such as the conspiratorial idea that it is a biologically engineered weapon created by the Chinese government to target America. These ideas often have little to no basis upon which they are built, yet those that come into contact with them - such as through social media or by word of mouth - may be negatively affected and develop a mindset of fear or unease towards factors of the pandemic that are related to them. Social behaviours play a large role in the development, transmission, and escalation of each of these ideas.

By extension, these ideas have also led to people performing actions as a result of their belief in them, whether it is done consciously or subconsciously. For example, one significant effect of the fear and panic surrounding the virus is the act of "panic buying", known as purchasing extremely large amounts of products such as food and toiletries in response to a disaster, which in this case is the COVID-19 pandemic. At the beginning of the pandemic, one of the products that became victim to this movement was toilet paper, and essential virus fighting products such as hand sanitizer, plastic gloves, and medical face masks were also hugely affected. This caused stocks to fall to an all time low and resulted in many lower-income families not having the means to access any of these products while those who were able to purchase them came to possess a surplus. Panic buying is considered a social behaviour because it is a phenomenon that came to be due to the response of the people who perceived COVID-19 as a threat to their daily lives. Psychological factors that may come into play to impact these actions include mental and emotional stress, pressure, and uncertainty, which all have the possibility of influencing individuals to take drastic and unnecessary measures as a coping mechanism.

In addition to the rise of panic buying, the increased incidents of racism towards those of Chinese and Asian backgrounds are also a result of the fear mongering surrounding COVID-19. Because the virus originated in Wuhan, China, the city eventually become known to be the epicentre of the pandemic, which led to an uptick in negative ideas and stigmas surrounding the Chinese population. These unfounded perceptions of Chinese individuals led to a stereotypical link of them to the COVID-19 virus, which then influenced a rise in racist behaviours towards those who were of similar ethnic origins. Avoidance of Chinese-owned businesses and restaurants was one of the key consequences of these ideas. Asian individuals also reported an increase in discrimination towards them or their friends and relatives - for example, following the Oscar victory of the South Korean film Parasite, an outburst of racist remarks comparing its title to the ongoing COVID-19 pandemic was recorded on social media sites and apps such as Twitter, Facebook, and Reddit. The conscious or subconscious association of the virus to Chinese and Asian individuals is a result of social behaviour being influenced by outside factors such as news and misinformation. These can lead to dangerous effects that manifest in the form of discriminatory words and actions.

Furthermore, in relation to the economical side of the pandemic's effects, a new rise in e-commerce and shopping through the Internet has begun as a result of the social distancing measures taken during the mandatory quarantine of most countries, which has been done in an attempt to prevent the rapid spread of the virus among the population. COVID-19 has impacted the social behaviours of consumers in a way that may leave a significant change in the way society operates. Due to the long-distanced remoteness of those who are under quarantine, many have taken to the Internet, using sites such as Amazon and eBay as a replacement for physical stores that they can no longer access because of the pandemic. In response to this, many large-scale corporations have also begun to welcome e-commerce and incorporate aspects of it into their own marketing and business models. COVID-19 has impacted the economy in a fundamental way, and the change in attitude towards e-commerce as well as the altered behaviours of its consumers may lead to a long-lasting shift in the dynamics of businesses and clients even after the pandemic has come to an eventual end.

The impact to the population's mental health also represents a key point of the change in social behaviours that the COVID-19 pandemic has brought to the world and international society. A variety of concerns have been raised in terms of the effects that an ongoing global pandemic may have on people, in particular vulnerable individuals such as youth, lower-income families, or those who are already suffering from pre-existing mental conditions, which has also led to a spike in resources and aid for the people who may be influenced by the darker side of the pandemic. Because COVID-19 has resulted in what is essentially social isolation, many individuals may find themselves becoming negatively affected by the reduced communication and interaction with the people they know and the world around them. The virus has also thrown much of society's regular functions, proceedings, and events into disarray, including any educational institutions and companies that have halted their activities for the time being while the pandemic continues. Additionally, people may feel that their levels of productivity and general enjoyment of their daily lives have dropped as a result of these extraordinary circumstances. All of these factors may play a part in the deterioration of individual mental health, which affects their social behaviours and may have the possibility of leading to long-term consequences on their well-being and state of mind.

Social behaviours and the way individuals interact with the people and events around them are an integral part of everyday life. As such, the COVID-19 pandemic has significantly impacted, changed, and altered how people perceive and communicate with their environment. Psychological factors may play a large role in affecting these behaviours, but ultimately, the international pandemic that has influenced virtually every aspect of daily life has done a great deal in terms of fundamentally adjusting and reshaping the way individuals function within their society, and many factors of these changes are likely to continue to be shown even after the pandemic eventually comes to an end.

CHAPTER 13

IMPACT OF COVID-19 ON THE CANADIAN ECONOMY

BY MIRAY MAHER

Introduction

On March 13th of 2020, the government of Canada issued a state of national emergency. As businesses shut down and borders closed, many were left without work. In addition, schools and child care facilities suspended as the health care system attempted to thwart the spread of COVID-19 and "flatten the curve". However, as some sectors faced a major downfall, others thrived and even exploited people's fears and anxieties going into a pandemic. Overall, this pandemic caused significant losses, with the Canadian GDP dropping 7.5% and 11.6% in March and April respectively, a record plummet. In addition, unemployment rates have been the highest in decades at 13.7% as of May, jumping from the average rate of 5.6% prior to lockdown (Statistics Canada). This is due to a large number of layoffs followed by a recession in multiple sectors such as manufacturing, hospitality and retail, which reduced their labour force to match the lowered demand. The government mitigated the worsening rates of unemployment by creating the Economic Response Plan which included benefits such as the Canadian Emergency Relief Benefit (CERB). To date, $64.72 billion has been paid to 8.5 million unique applicants, allowing the country to lockdown non-essential businesses while supporting those affected by the shutdowns by enabling them to put food on the table amidst this uncertainty (Statistics Canada). In this chapter, we will discuss the impact of the pandemic on key industries of the Canadian economy like tourism and travel, retail and real estate market.

Travel Industry Takes the first hit

When the federal government declared a national state of emergency, international borders closed to non-essential travel and even interprovince travel was restricted in Prince Edward Island, New Brunswick, Nova Scotia, and the Northwest Territories (Collie). In fact, the number one Canadian airline, Air Canada, has reported a $1.752 billion net quarterly loss, laid off thousands of workers, and cut seating capacity by over 90% (MacGregor). Furthermore, the worsening state of the United States' COVID-19 situation drives the continued closure of the U.S-Canada border. This is exacerbating the economic losses tied to revenue from American tourists, which constituted almost 69% of all tourists coming from abroad. Over two-thirds of American visitors arrived by car and 40% made same-day trips that usually involve shopping and sightseeing, contributing significantly to the Canadian economy (Statistics Canada).

This sudden stoppage of all travel halted all revenue targeted towards tourism as well. According to the Tourism Industry Association of Canada, the tourism industry accounts for over $100 billion of annual revenue and 1.8 million jobs across the country, allowing 1 out of every 11 jobs to be in this sector by the end of 2018 ("How Canada's Tourism Industry Is Trying to Salvage Summer 2020."). Unfortunately, the COVID-19 pandemic drove 28% of those in the tourism industry to unemployment by the end of April (Hensley). With the suspension of leisurely travel, and the cancellation of sports events, conferences and concerts, hotels have been seeing a steep decline in economic activity, with hotels only holding 10% of its usual occupancy on average, causing more than 4,100 hotels to close and more than 250,000 employees to be laid off all over Canada. Toronto, for example, has experienced a loss of over $250 million in revenue from April through June in cancellations of major events. Montreal predicts a loss of over $55 million in sales with the added loss of over 70,000 cancelled hotel bookings (Reynolds). Banff National Park, the major tourist attraction of Alberta, contributes $3 billion of Alberta's $9 billion tourism economic activity. With almost half of Banff's economy driven by tourism and half of the visitors coming from abroad, the COVID-19 crisis took a huge toll on the small town of Banff as 6000 of its small 9000 population were laid off. Furthermore, the pandemic devastated the small province of Prince Edward Island, which relies on tourism to generate almost $505 million in annual revenue as tourism is one of the top three economic sectors of the island (Hensley). Although the hospitality sectors was hit first due to the cancellations of many events and bookings, employees of all sectors are economically impacted. Small businesses especially take the fall of this ripple-like effect as explained by the vice-president of destination development of Tourism Toronto, Andrew Weir: "Obviously, hotels and restaurants are affected to their very core. But the reality is when meetings and conventions don't happen, that affects audio-visual companies, it affects transportation companies, it affects staging companies and the entertainers,

98

the musical acts and professional M.C.s that are hired to host these events" (Reynolds). This gives a glimpse of the extent of the financial damage that the COVID-19 crisis has caused and the wide scope of businesses it has affected.

Shopping Patterns During COVID-19

Unemployment, financial difficulty, and the fear of infection are large factors that trickle down to consumer patterns, negatively affecting some businesses but also significantly profiting others. Overall, Canadian household spending has dropped by 2.3%. This could be explained by income loss, job insecurity and lack of opportunities to spend due to the closures of many non-essential businesses such as restaurants, retail stores and recreational services. This drop occurred in spending on goods and services, with a drop of 1.7% and 2.8% respectively. Durable goods fell substantially by 6.4%. Goods in this category include cars and other types of vehicles. This is explained by income uncertainty and a lesser need for transportation vehicles as working from home became more common. Spending on semi-durable goods such as clothes and shoes also fell by 9.4%. Lastly, non-durable goods such as food and beverages rose 3.1% reflecting panicked consumer behaviour as they stockpiled these goods, especially non-perishables such as canned beans and vegetables, pasta, rice and flour (Statistics Canada). This increase in grocery sales offsets the decrease in economic activity in restaurant and hospitality establishments as people steered toward cooking and eating at home more often. Certain products such as bikes have also been in very high demand as people try to find healthy and safe recreational activities during the time of social distancing (Evans).

Furthermore, as people stayed more indoors, more money was allocated to leisure and entertainment. Books, movies, and music increased in sales, with 22% of Canadians reporting buying more books in March than in February (Guldimann; Hirchberg).

Retail stores have not been fairing well during the pandemic, with an estimated loss in profits of almost 25% in April, the biggest drop yet. Grocery stores deemed as an essential establishment saw a rise in sales, but shops that sold less compulsory items, were forced to close. The stores that saw the biggest plunges in sales were car dealers (-44%), gas stations (-32%), clothing and accessories stores (-70%), and furniture stores (-51%) (Evans). As people avoided going in public to prevent catching the novel coronavirus, many relied on online shopping to buy products and services and this left businesses without a strong online platform to sell their products with a more substantial financial loss as fewer people had access to their goods. Businesses were forced to rapidly drive the creation of an online shopping platform to stay competitive in the market. Businesses with a strong eCommerce presence thrived. In fact, online shopping increased by over 120% during COVID-19 lockdown and online shopping purchases made up 10% of everything sold in the country. This figure does not include the popular American eCommerce company, Amazon, as it is still considered foreign-based. An example of this is Shopify, an Ottawa-based Canadian company that is valued at over $140 billion that helps stores and businesses set up an online platform to sell their products. Such services have been in critical demand as the whole world shifted digitally (Evans).

COVID-19 has also shaped Canadians' payment preferences. Canadians have moved away from cash and cheques and more towards contactless payment methods. 62% have reported using less cash and 40% of weekly cheque users have reported a decrease in their use. Canadians are more frequently using Interac e-transfer and PayPal. In fact, 42% of Canadians surveyed admitted to avoiding shopping at places that did not offer contactless forms of payment ("COVID-19 Pandemic Dramatically Shifts Canadians' Spending Habits.").

Housing Market

The housing market is another aspect of the economy that directly impacts the consumer choices of Canadians. Many are worried that this pandemic would alter the housing market in a way unfavourable to them. Those who were retiring and had planned to use the equity built from their houses for their pension are worried that falling housing prices would decrease the amount of money they could attain. Others hoped

100

that dropping prices would allow them to finally afford to buy a home.

At the beginning of the national emergency state, unemployment and the pause in incoming new immigrants significantly decreased housing sales by 70% (Langton). However, that did not significantly impact housing prices as demand and supply decreased roughly by the same amount (Borzykowski). As stated on StatsCan, "Demand outpaced supply in most key housing markets until mid-March 2020, spurred by economic growth, a low unemployment rate and population increase largely from immigration." According to Scotiabank's estimates, the housing prices are expected to drop by a mere 4% from 2019, a drop that will easily be overcome and even exceeded by mid-2021. It is worthy to note that this drop is much lower than what would have been anticipated due to the Canadian government supporting Canadians through various benefit programs such as the CERB (Borzykowski). The biggest risk to the housing market currently is the decline in immigration as the new immigrants drive new house sales.

The Path to Recovery

As tourism took a huge fall, the Canadian government looked to support and promote domestic travel. The Canadian government has announced that $30 million is going to be spent supporting provinces and territories to promote vacations for domestic residents that are looking for a nearby getaway. Ben Cowan-Dewar, the chairman of Destination Canada said in a statement that the aim of this government initiative is to provide a "valuable lifeline" for this important and struggling sector during its peak summer season. Many Canadians who would have usually opted for trips to Europe or the Caribbean have now chosen to look for something closer to home due to restrictions to air travel, a positive outcome for Canadian establishments (Hensley).

Moreover, as the restrictions slowly loosen up in stages, businesses can finally open their doors once more. This will allow economic activity to resume and the circulation of money will open the opportunities for employment for Canadians. Those who will see the largest bounce-back as restriction are the sectors that produce goods (ex. factory production lines). This will be followed by face-to-face services, where there is greater risk of viral transmission. Businesses that go under this category are hair salons, massage parlours and dental offices. The entertainment industry may not return to normal soon because of concerns about large gatherings and close contact. This includes concerts, movie theatres and amusement parks (Schembri).

Taking Ontario as an example, the province has been very careful in its economic recovery plans and have lifted restrictions controllably in stages with each stage lasting 2-4 weeks. Stage 1 involved reopening stores with a separate street entrance, motor vehicle dealerships, facilities that conduct life science research, and some outdoor seasonal activities like golfing ranges and cycling tracks. In stage 2, places of worship were allowed to open but with a cap of 30% of building capacity. In addition, establishments providing personal care services such as hair salons, tattoo parlours, and spas were allowed to reopen with the exception of services that cater to the face. Facials, face tattoos, and eyebrow grooming remained prohibited. The latest stage, allows nearly all businesses and establishments to open so long as the proper safety protocols are adhered to. However, these high-risk places have been exempted from reopening: movie theatres, buffet restaurants, and amusement and water parks ("Reopening Ontario").

It is important to note that within all these stages, physical distancing and the donning of a non-medical face covering were all still enforced.

The controlled reopening of the economy ensured that millions of people could safely return to work while preventing the occurrence of the second wave. This best ensures that these reopening stay permanent, as a second wave would prompt the closing of various businesses and establishments again.

Conclusion

It is by no surprise that COVID-19 shook many aspects of everyday life. Arguably, the largest reaching effect was the closing of many parts of the

Canadian economy. Many non-essential businesses shut down and many lost their jobs or at the very least, had a cutdown of their hours. To alleviate the losses of income, the Canadian government put money back in the citizens pocket to afford basic necessities among this crisis. Although there was a major decline in the travel and tourism industry, the government implemented a recovery plan to support domestic tourism. Furthermore, the pandemic caused a shift within consumer patterns such as shopping online more, preferring contactless methods of payment, and increase in the purchases of non-durable goods. Lastly, the housing market, although endured a small decline, is expected to bounce back even stronger within the next year. As Canada continues on its path of economic recovery, the main priority will be keeping new COVID-19 cases to a minimum, protecting the most vulnerable and maintaining a high standard of public health safety.

CHAPTER 14
LONG TERM SOCIAL IMPACT OF COVID-19

BY GRACE LETHBRIDGE

The COVID-19 pandemic impacts all members of the population; however, it is particularly detrimental to members of vulnerable groups, such as people living in poverty, older persons, persons with disabilities, and indigenous peoples. People without access to running water, refugees, migrants, or displaced persons also suffer disproportionately from the pandemic and its aftermath — whether due to limited movement, fewer employment opportunities, increased xenophobia, etc. Additionally, several social issues such as food security, domestic violence, and social isolation are being exacerbated, and social inequities continue to be highlighted as a result of current stressors. If the government does not initiate new long-term public policies to address the social crisis caused by the pandemic and its aftermath, issues such as inequality, exclusion, discrimination, and global unemployment may be further deepened within these networks of societies. Comprehensive social protection systems that provide basic income security are needed in order to protect workers, reduce the prevalence of poverty, and enhance people's ability to manage and overcome shocks. This chapter will focus on the emerging social issues within the workforce, education system, and community at large.

Workforce

In March 2020, the Canadian government enacted policies to contain the COVID-19 virus. These policies and changes caused abrupt disruptions to many businesses across Canada. Social distancing, border control, and quarantine measures all caused unforeseen challenges within the workplace. These challenges caused many small businesses to close permanently, as they were unable to generate enough profits to survive the pandemic. Other large businesses such as David's Tea, Starbucks, GNC, J. Crew, etc. were also impacted by the pandemic, resulting in bankruptcy or permanent closures of certain stores. As a result of financial loss, many businesses laid off their employees. The Canadian government, through their Labour Force Survey, has been studying employment trends by collecting information during a single week each month and classifying the Canadian population aged 15 and older as either employed, unemployed, or not in the labour force. Over

3.1 million Canadians were affected by either job loss or reduced hours over the time change of February to March 2020(Government of Canada). Unfortunately, even businesses that are not severely impacted by the pandemic, have still used this time as an opportunity for company pruning to save resources and cut costs. With high unemployment rates and decreased employment opportunities, many Canadians are burdened with financial struggles, further mobilizing the current social inequities within our society. Although individual companies may justify these actions by attributing them to the pandemic, companies should be accountable for their divisive measures during a time of crisis. Jobs are more than just the way someone makes a living; they influence peoples' identities. Many jobs offer social outlets, which give employees structure, purpose, and meaning to their lives. Losing a job, aside from the obvious financial impact on one's life, can impact one's mental health as they try to navigate new uncertainties within their life.

To overcome social distancing measures, many businesses have converted their operations to online platforms. With the successful conversion to operating online, many businesses have noticed that they can reduce expenses by continuing to operate entirely online indefinitely. Although this may provide employees with more flexibility in their life, many people thrive in social environments where they can develop relationships with their co-workers. This type of social interaction may be difficult to mimic in online settings, which could consequently negatively impact group processes. Team psychological safety has been defined as an atmosphere within a team where individuals feel comfortable engaging in discussion and reflection without fear or censure. Team members are thus more likely to hold productive discussions through asking questions, seeking feedback, highlighting failures and sharing information because their focus is not on self-protection(Edmondson and Lei). Permanent online workplaces may sacrifice a company's ability to cultivate team psychological safety. It will be imperative that businesses address this social aspect of the workplace as they adapt to new online environments.

Education

Much of the planning has focused on the level of community spread to determine if schools should restart, and if so, how they should restart. To curb the spread of COVID-19, a combination of masking, physical distancing of two metres, and practising proper hand hygiene will be required. Many schools around the world are not equipped to implement all of these measures. In particular, creating two metres of physical distance in classrooms would require more resources, both physical and human, than most jurisdictions appear to be willing to invest in. The Ontario back to school plan has received much criticism from teachers, parents, and health professionals. One of the biggest concerns is not reducing class sizes in elementary schools. Class sizes are typically 20 to 29 students per room. These densities will not allow for proper physical spacing of two metres between students. Although young children with no comorbidities are considered to be at a lower risk of serious health outcomes, there are concerns that outbreaks in schools pose a more serious risk to students' families, teachers and other staff, and their families. There are parents in Ontario, and in many other jurisdictions, who are planning for their children to participate in online classes. This is not feasible for many families in lower socio-economic classes (e.g. families that require two parents to work, single parent families, and parents who do not have the option to work from home). Also, access to high-speed internet is not universal due to the cost associated with it or the unavailability in remote regions. There is concern regarding how the government will fund public education in the future. In an effort to reduce their investment, governments may use the pandemic as a means to push for more online classes, which are deemed to be less effective than in-person classes. This is particularly concerning in Ontario because the government announced their plans to implement mandatory online credits for high school students prior to COVID-19. If public education does not receive appropriate funding, more families will move their children to private schools, creating further community segregation. The resulting segregation of students by wealth will only further perpetuate a system of inequity. This could lead to underfunded public schools, thereby disadvantaging the children of lower income families.

Even for families who can and will choose to have their children learn online, there are concerns regarding the negative impacts of isolation from peers and teachers on the mental health of their child. Although online learning may be sufficient for some students, but not for the majority. It will be particularly difficult for learners who need face-to-face interaction and a sense of community in order to learn.

Students attending in-person classes may experience negative impacts resulting from wearing masks and physically distancing. The curriculum in the primary grades involves tactile learning and the development of fine motor skills. The long-term impact of these safety measures to prevent the spread of COVID-19 on childhood development remains unknown.

Family Dynamics: Domestic Violence

Stressors caused by the pandemic, such as job losses, social isolation, health concerns, school closures, mental health challenges, and addictions are adding pressure and increasing domestic violence, which includes emotional, physical, and sexual abuse (United Nations). With lockdown measures enforced, many women are trapped at home with their abusers, struggling to access services that are currently experiencing financial cuts and restrictions. Quarantines and social isolation mean that abusers and victims are living in close proximity for extended periods of time, and other people are not present to notice the signs of violence and intervene.

A survey completed by Statistics Canada reported 1 in 10 Canadians are concerned about the possibility of domestic violence as a result of COVID-19 (Government of Canada, The Daily — Canadian Perspectives Survey Series 1). During the pandemic, everyone has been told to stay at home to remain safe; however, for those experiencing domestic violence, this could not be further from the truth.

Food Insecurity

The Canadian Mental Health Association (CMHA) reported 21% of Canadian parents with children under 18 years of age are worried about having enough to feed their families (CMHA). As more individuals and families experience the impacts of financial vulnerability, the cascading effects of food insecurity will be felt by vulnerable communities. The closures of many food banks, daycares, and schools as a result of the pandemic, coupled with financial insecurity, further exacerbates food insecurities. Many individuals and families are simply concerned about having enough food, let alone nutritious food. The Canadian government recommends eating a well-balanced diet, with an abundance of fresh fruits and vegetables in order to maintain a healthy immune system. They also recommend minimizing trips to the grocery store. Both of these recommendations come at a significant cost. Fresh fruits and vegetables may be nutritionally dense, but they are not calorically dense. Therefore, you would need to eat a lot more fruits and vegetables than a cheap frozen pizza to satisfy your hunger. Many people with financial struggles are unable to afford the luxury of fruits and vegetables and are instead forced to buy cheap and highly processed foods. Additionally, not everyone can afford to reduce their trips to the grocery store. Many individuals and families who lack financial stability, and are living from paycheck to paycheck, are unable to afford two weeks' worth of groceries in one trip. Financial instability is a key contributor to food insecurity, which is why it is imperative the government understands and addresses the interconnection of these social inequities.

Mental Health and Addiction

Mental health and addictions are highly complex and contextual to each individual and their unique situation. Factors contributing to poor mental health include experiences of trauma and abuse, inaccessibility to jobs, education, insufficient housing, social isolation, lack of community, and food insecurity (Public Health Agency of Canada). The COVID-19 pandemic has further magnified these factors, leading to increased mental health issues. The mental health implications of the pandemic will likely last longer and affect a larger percentage of the population than the pandemic itself (Ornell et al.). The pandemic is a traumatic event experienced by the collective. The varying reactions and levels of resiliency in response to the traumatic effects of COVID-19 are likely linked to degrees of privilege. As a result of the pandemic, 50% of Canadians have reported a worsening of their mental health (Government of Canada). The primary feelings reported are worried, anxious, and bored. Not only can prolonged periods of social isolation lead to mental health issues, but these extended periods of alone time give individuals time to ponder these negative thoughts.

Mental health and addiction are strongly correlated to each other. It is estimated 50-75% of people who experience addiction also suffer from one or more mental health disorders (Marshall et al.). With mental health decline, individuals may look to substances as a coping mechanism.

Social Protection Systems

Regional, provincial, and national governments are responding rapidly with social policies to the economic impacts of COVID-19 by establishing individual benefits, wage subsidies, business sector supports, grant programs, tax payment deferral options, and other social policies. These policies strive to mitigate immediate impacts; however, since these supports are time-limited, it could result in increased social issues relating to housing, food security, and mental health when the deferrals or benefits stop.

CHAPTER 15
IMPACTS ON THE PUBLIC HEALTH SYSTEM

BY LILY LIU

The COVID-19 virus has led to an international pandemic affecting virtually every country in the world. First identified in Wuhan, China in December 2019, the highly infectious novel coronavirus disease has seen a rapid spread within the global population and caused over 20 million cases of infected individuals and more than 750,000 deaths across 188 countries as of August 2020. Because of its infective nature and the speed of its potential transmissions, the virus has become an extraordinary issue that influences close to every aspect of everyday life, from the world economy to the complex system of global politics as well as the fundamental structure of human society.

The healthcare system, both public and private, can perhaps be referred to as the social sector that has been the most impacted in the ongoing COVID-19 pandemic. The drastic increase in cases around the world - which comes with the swift transmission of the virus from patient to patient - has resulted in medical care institutes such as hospitals becoming vastly overwhelmed by the volumes of infected patients, people seeking COVID-19 tests, and individuals self-isolating after exposure to infected patients or after travelling abroad in countries affected by the pandemic. The highly contagious and easily transmissible nature of the COVID-19 virus has resulted in an extraordinary spike in cases, which, by extension, leads to a dramatic rise in patients requiring all types of health care, medical assistance, and medical attention. This has led to enormous stress and pressure on the public health system as a whole.

At the surface of the current issues around the coronavirus disease, public health systems have been subject to a surge in patients and demands from all sides of the pandemic. Staff shortages have become a common problem among hospitals, particularly among third-world countries as well as areas that are considered fairly low-population or not as advanced in comparison to other locations. Critical shortages in crucial emergency supplies such as medical masks and hand sanitizer - as well as more healthcare-focused items used in hospitals such as ventilators and protective equipment for its staff members - have also been reported across the world. These supply shortages, all of which are key parts of the global effort towards fighting the COVID-19 pandemic, have caused inadequate treatment of patients and have led to increased risks in the virus transmission among staff members and other visitors to the hospital.

Although these supply shortages may be considered short-term impacts as opposed to more serious and lasting long-term impacts, they possess a huge role in terms of escalating the pandemic and causing more cases to be confirmed, as a lack of proper equipment results in insufficient protection towards the individuals involved in the public health system. A high number of doctors and nurses treating patients for the COVID-19 have tested positive for the virus for this reason; in the absence of needed supplies, the possibility of being infected with the virus within hospitals and other medical care spaces will only grow larger due to the widespread exposure to patients who are positive cases and have already been infected.

Additionally, the mass scale of the current pandemic surrounding COVID-19 have also caused disruptions in regular medical procedures and health care meant for other diseases such as chronic and terminal illnesses, whose patients may receive less satisfactory treatment due to the focus and attention paid to the COVID-19 virus. Shortages in required supplies and equipment may deeply impact these patients as well, some of whom will be more vulnerable to COVID-19 transmission due to their already compromised immune system. All of these impacts to the public health system have the risk of heavily affecting its staff, its general methods, and the way medical institutions are handling the entirety of the ongoing COVID-19 pandemic.

Furthermore, the first confirmed cases of COVID-19 in Canada can be traced back to long-term care retirement homes that house seniors, as was the case with the province of British Columbia. Multiple other long-term care homes in Ontario also became locations where there were a number of confirmed cases, largely due to its vulnerable inhabitants, who are senior citizens with weaker immune systems and less means of defense against the COVID-19 virus. Much like hospitals, these homes may suffer a sharp decrease in needed supplies and equipment and a stark increase in infected staff and patients as a result of their exposure and isolation with patients who have contracted the virus. Although long-term care retirement homes are not necessarily considered part of the public health system, they are inherently linked to it due to their close relations with medical facilities such as hospitals as well as their housing of seniors who are often heavily compromised in terms of health and immunity.

In terms of the long-term impacts that the COVID-19 pandemic may have on the state of the world in the future, there are many theories surrounding the nature of the virus and whether or not it is possible for patients to be re-infected. In February 2020, the government of Japan reported a confirmed case of re-infection in a woman a middle-aged woman who had tested positive for the second time after her previous recovery from the virus, which closely followed another similar case of COVID-19 re-infection that had been reported by China earlier within the month. This managed to raise several international concerns on the possibility of the virus bypassing the usual immunity barrier of patients who have already been infected to infect them again.

However, medical experts have pointed out that in these types of unique situations, it is fairly difficult to determine if the patient was truly infected with the virus again, or if they had never recovered the first time and merely became ill again as a result of the lingering remnants of the virus that had not left their bodies. Because COVID-19 is a novel coronavirus that had emerged only recently, there is little information available on its specific details and properties, and as such, the theory that it is able to re-infect select patients may not be able to be confirmed.

If the virus is capable of re-infecting previously healed patients, however, then there raises an issue surrounding the immunity guarding against COVID-19. In this case, there may be a situation in which the currently ongoing pandemic becomes a seemingly never-ending endemic virus that returns in seasonal periods each year, much in the way that influenza - or, as it is more commonly referred to, the flu - is already. Furthermore, even if the COVID-19 virus does not possess these re-infecting properties that are being theorized in the present, it still poses a risk of returning as an endemic virus with yearly outbreaks. Four of the other coronaviruses that affect humans - one of which is the virus that causes the seasonal flu - have already become seasonal, which gives way to a possibility of COVID-19 following the same path, especially due to the lack of immunity against it among the human population and the current absence of a functioning virus.

Should this become reality, then the public health system would suffer greatly as a result. As it stands, the current state of international society is still ill-equipped to deal with the presence of a highly infective and easily spread novel coronavirus, which has led to a severe shortage of medical staff and supplies and an overwhelming of the healthcare system around the globe. The impact that the COVID-19 pandemic has had on the system in the present would be likely to multiply and repeat each year if the virus becomes endemic. This would hold drastic consequences, as many portions of the world - such as the economy in particular - are currently already weakened due to the rapid spread of the pandemic.

The public health system may be forced to go through many more periods of difficulty in the future, not only due to the potential recurring seasonality of the COVID-19 virus, but also due to the weaknesses and fractured points in the parts of society that are inherently connected to both public and private healthcare. Under the current circumstances of the world, the best course of action to take may be to simply continue to study and learn from the current pandemic, and then adequately adjust methods of prevention and protection for the future.

CHAPTER 16
CONCLUSION

BY MIRAY MAHER AND TINA WU

After the ever-expanding world of zoonotic viruses crossed paths with the human population in December 2019, COVID-19 has infected millions of people globally and touched upon all aspects of the human experience. The pandemic has caused many political systems to take action to contain the highly infectious virus. The effects of both the virus itself and the newly enacted policies in response have major economic, social, and cultural implications. As funding is allocated towards SARS-CoV-2 research, many fields of science are applied to understand the workings of the novel coronavirus. From weather patterns and the environment, viral characteristics of SARS-CoV-2, the evolution of coronaviruses in animals, and coronavirus symptomatology, we have looked at the virus through a variety of lenses to uncovered pieces of the truth regarding its rapid spread.

As uncontrollable as COVID-19 seemed at the beginning of this pandemic, scientific research became a beacon of hope in the midst of the panic. From the biological perspective on SARS-CoV-2, a lot has been learned. Examinations of environmental patterns that lead to COVID-19 spread can help us determine the policies that need to be implemented to combat the virus. Since poor air quality and circulation can perpetuate COVID-19 spread, policies have been enacted to close down areas of large social gatherings, particularly indoors. Even while stores are reopening in Canada, it is commonplace for businesses to hand out masks at the storefront, encourage keeping a six feet distance from other shoppers, and provide hand sanitizer for use due to the higher risk of disease spread in a closed indoor space. Unfortunately, despite many Canadians hopeful that warmer summer temperatures will bring down the rates of coronavirus transmission, research has found that COVID-19 viral particles covered in mucus and saliva are resistant to extreme temperature conditions and the effects of climate change might actually make viruses more resistant to the heat. On the brighter side, this implies that the decrease in Canadian cases since May could be the result of policies and behavioural changes, such as social distancing and limiting group contact, rather than seasonal change. Additionally, a study of environmental factors on coronavirus spread indicates a need for greater regulation on animal husbandry practices. To prevent farms and wet markets from becoming a breeding ground for viruses, there needs to be greater education on the types of practices that can help prevent viral transmission.

Environmental patterns are only half of the picture. The infectivity of SARS-CoV-2 can also be described by its viral characteristics. Research in this field is extrapolated to find ways to curb the virus's high transmission rate, by developing therapeutics that target those viral characteristics. The prominent spike proteins of coronaviruses are a focus of research, as they are key to viral attachment and entry into the cells in the lungs. Contrasting SARS-CoV-2 with other coronaviruses is another method of determining viral characteristics that contribute to infectivity. But as much as we have learned about the structure of the novel coronavirus, it is important to stay humble.

By studying the evolutionary history of coronaviruses, it is clear that a lot is left unknown. Scientists continue to piece together the past to learn more about the coronavirus ancestor and its diversification into the various coronaviruses present today. Regardless of whether the ancestor appeared 10,000 or a million years ago and how it branched off, an exploration of viral evolution and mutation allows for a better understanding of how a virus that is normally transmitted between animals can spill over into the human population. Additional study of the symptoms associated with coronaviruses in animals and humans alike can help us describe the body's response to viral invasion and the potential effects that new coronaviruses may have on our physiology. Furthermore, by exploring the viral mechanisms of infections in animals, it becomes more apparent which animal species carry viruses that are at risk of spreading to humans, allowing policymakers to implement the necessary protocols to prevent transmission of new viruses to humans. Such policies would need to consider the various socioeconomic and cultural factors contributing to behaviours conducive to viral spread. Poverty drives populations to practice unsafe behaviours that could incite epidemics and pandemics such as this. An example is building infrastructure close to wild habitats and deforestation. While the idea of another pandemic appearing in our midst is horrendous, we cannot shy away from this possibility. It is tempting to cut funding for research as soon as the threat is over, but if we have learned anything from the SARS epidemic in 2003 and our COVID-19 pandemic now, it is that no research is wasted. Any additional information, resources, or interventions that could have been gained had the research on SARS continued would have been astronomically valuable now that a similar coronavirus has infected millions of people worldwide. It is too late to return to the past and make changes. However, it is never too late to change our ways so the world will never have to live through a pandemic to this scale again.

With all that COVID-19 has contributed to scientific discovery, it has an even greater influence on media, politics, economics, and society. It is paramount to consider the government policies enacted in response to the pandemic. If a similar situation were to happen many years into the future, the greater population must know what worked to combat COVID-19 spread and what policies were ineffective. In addition, they must closely inspect the economic effects of both the virus and the government policies to determine whether the benefit of the policies enacted is worth the loss of employment and the billions of dollars spent on government assistance programs while risking a major economic recession. As air travel became restricted, Canada lost a major source of revenue from tourism and hospitality services. Shopping and retail also shifted in favour of online substitutes as people refrained from going out and found convenience in getting the essentials at the click of a button. If work-from-home was not an option, many individuals gave up their jobs. Evidence shows that there is a link between unemployment and depression ("Unemployment and Mental Health"). Governments must make a hard decision in balancing between the economic and mental well-being of citizens, and the demands of the public health emergency. The approach in Canada was to increase public spending and to give out Canadian Emergency Response Benefits for individuals who have lost their jobs. Other aspects of consumerism give insight into the public reaction to COVID-19. Contactless methods of payments became the norm, with cash and cheque taking a step back. The change in consumer patterns during the pandemic due to both the coronavirus and government policy may be here to stay, shaping the way we go about our daily lives years after this crisis is over.

The social implications of COVID-19 policies must also be examined to determine the cost-benefit. School closures have a large impact on young children especially those who need social and environmental stimulation to fully grow and develop. Learning from a screen is simply not the same as the in-classroom experience. It can be difficult for elementary school students to learn crucial and foundational concepts at home, where there may be no adult aid to substitute a teacher's. Furthermore, adolescence is a time where social relationships begin to bloom as young children discover themselves and their interests. This sparks the question of whether children, who seem to be less susceptible to COVID-19 than adults (Bloomberg), would benefit more from staying at home or from attending schools. This consideration has schools have decided to reopen in the fall to some extent for many students; some schools have opted for a hybrid method, alternating between online and in-person learning and others might have full or shortened school days. The educational impact of COVID-19 shows the need for better remote learning protocols in the future and innovative solutions to engage students in their learning, even when the classroom is far away.

Socializing in present-day society has also taken a turn. Many have opted for social media in place of physical interaction, with group chats and calls as the new "going out". As internet and social media usage spikes, so does false information. A flashy title and a shocking message can spread far as social media algorithms spread the most popular images, videos, and information. The greater the reaction a post obtains, the more people the message will reach. The effects of false information are not as benign as one may think. Perpetuated stereotypes have harmful effects against targeted individuals or the general public as a whole. One of these effects is racist attitudes towards East Asians. When powerful leaders perpetuate that COVID-19 is a "Chinese virus" it may incite violence and discrimination against Asian-American individuals (Donaghue). Since the pandemic, there has been a severe increase in incidents of xenophobia and racism against this demographic. Furthermore, a general distrust towards science and health officials has left many protesting the wearing of a non-medical face-covering and social distancing, asserting that it is an infringement upon their rights. Some even go to the extent of claiming that masks are harmful, comparing it to a muzzle. Myths about masks depriving the body of oxygen or leading to carbon dioxide toxicity also flood the internet without any scientific backing whatsoever. False information often clouds facts, leaving many susceptible, especially those who do not come from a scientific background. This goes to show the constant need for digestible scientific literature surrounding this topic to constantly engage and inform the largest number of people possible. The flood of false information may call for action to educate populations on discerning fact from fiction in an online setting. It also brings to light questions on whether or not social media platforms should intervene when false messages are conveyed, and if that would be considered a violation of freedom of speech.

Comparing and contrasting government responses is another way to identify which responses to COVID-19 were effective. With China, we could see that the coordinate and crush method with unprecedented lockdowns in the province of Hubei was successful in combating viral spread, although enacted too late due to government attempts to cover up an early alert on the disease. In Canada, we can see that physical distancing policies and other measures like school and business closures have appeared to ease the slope of coronavirus cases. In the United States, it is apparent that vaccine research is progressing, but the government's response to delaying viral transmission is heavily criticized.

The effects of COVID-19 are far-reaching. The pandemic has devastated the economy, caused extensive panic, and jolted the world into action. But we are not without hope. If we take inspiration from our successes and learn from our mistakes, humanity can progress into a future where pandemics like these are only a dark flicker of the past.

WORKS CITED

Works Cited (Chapter 1)

Brown and Moffit. "What Happens When There Is A Pandemic? | CORONAVIRUS" YouTube, uploaded by AsapSCIENCE, 4 Mar. 2020, https://www.youtube.com/watch?v=oqtfqVsFaqc

Fox, Maggie. "Pangolins May Have Incubated the Novel Coronavirus, Gene Study Shows." Coronavirus, CTV News, 31 May 2020, www.ctvnews.ca/health/coronavirus/pangolins-may-have-incubated-the-novel-coronavirus-gene-study-shows-1.4961699.

Ma, Yueling, et al. "Effects of Temperature Variation and Humidity on the Death of COVID-19 in Wuhan, China." Science of The Total Environment, vol. 724, 1 July 2020, p. 138226., doi:10.1016/j.scitotenv.2020.138226.

Saif, Linda J. "Bovine Respiratory Coronavirus." Veterinary Clinics of North America: Food Animal Practice, vol. 26, no. 2, 1 July 2010, pp. 349–364., doi:10.1016/j.cvfa.2010.04.005.

Shi, Jianzhong, et al. "Susceptibility of Ferrets, Cats, Dogs, and Other Domesticated Animals to SARS–Coronavirus 2." Science, vol. 368, no. 6494, 29 May 2020, pp. 1016–1020., doi:10.1126/science.abb7015.

Wu, Katherine J. "There Are More Viruses than Stars in the Universe. Why Do Only Some Infect Us?" National Geographic, 15 Apr. 2020, www.nationalgeographic.com/science/2020/04/factors-allow-viruses-infect-humans-coronavirus/.

"The Classical Definition of a Pandemic Is Not Elusive." World Health Organization, World Health Organization, 1 July 2011, www.who.int/bulletin/volumes/89/7/11-088815/en/.

Corona, Angel. "Disease Eradication: What Does It Take to Wipe out a Disease?" ASM.org, 6 Mar. 2020, asm.org/Articles/2020/March/

Disease-Eradication-What-Does-It-Take-to-Wipe-out.

"Coronavirus." Centers for Disease Control and Prevention, Centers for
Disease Control and Prevention, 15 Feb. 2020, www.cdc.gov/
coronavirus/types.html.

"Does Polio Still Exist? Is It Curable?" World Health Organization, World
Health Organization, 20 Jan. 2020, www.who.int/news-room/q-a-
detail/does-polio-still-exist-is-it-curable.

Hegarty, Stephanie. "The Chinese Doctor Who Tried to Warn Others about
Coronavirus." BBC News, BBC, 6 Feb. 2020, www.bbc.com/news/
world-asia-china-51364382.

"Q&A On Coronaviruses (COVID-19)." World Health Organization, World
Health Organization, 17 Apr. 2020, www.who.int/emergencies/
diseases/novel-coronavirus-2019/question-and-answers-hub/q-a-
detail/q-a-coronaviruses.

Schumaker, Erin. "Timeline: How Coronavirus Got Started." ABC News, ABC
News Network, 28 July 2020, abcnews.go.com/Health/timeline-
coronavirus-started/story?id=69435165.

"Sick Leave." Ontario.ca, 3 Apr. 2020, www.ontario.ca/document/your-guide-
employment-standards-act-0/sick-leave.
" Canada." Worldometer, 10 Aug. 2020, www.worldometers.info/coronavirus/
country/canada/.

Works Cited (Chapter 2)

"About Middle East Respiratory Syndrome (MERS)". Centers for Disease
Control and Prevention, 2019. https://www.cdc.gov/coronavirus/
mers/about/index.html

"Influenza (flu) - Symptoms and Causes". Mayo Clinic, 2019. https://www.mayoclinic.org/diseases-conditions/flu/symptoms-causes/syc-20351719

"Middle East Respiratory Syndrome Coronavirus (MERS-CoV)". World Health Organization, 2019.
https://www.who.int/emergencies/mers-cov/en/

"Severe Acute Respiratory Syndrome (SARS)". Centers for Disease Control and Prevention, 2017.
https://www.cdc.gov/sars/about/fs-sars.html

"Types of Influenza Viruses". Centers for Disease Control and Prevention, 2019. https://www.cdc.gov/flu/about/viruses/types.htm

Groth, Leah. "Coronavirus Symptoms VS Cold: How Do They Compare?". Explore Health, 2020.
https://www.health.com/condition/infectious-diseases/coronavirus/coronavirus-symptoms-vs-cold

Joseph, Andrew. "WHO: Coronavirus is different from influenza, and that means it can be contained". STAT News, 2020. https://www.statnews.com/2020/03/03/who-coronavirus-different-than-influenza-can-be-contained/

Olena, Abby. "Scientists Compare Novel Coronavirus to SARS and MERS Viruses". The Scientist, 2020. https://www.the-scientist.com/news-opinion/scientists-compare-novel-coronavirus-to-sars-and-mers-viruses-67088

Ries, Julia. "Here's How COVID-19 Compares to Past Outbreaks". Healthline, 2020. https://www.healthline.com/health-news/how-deadly-is-the-coronavirus-compared-to-past-outbreaks#Seasonal-flu

Seladi-Schulman, Jill. "COVID-19 VS. SARS: How Do They Compare?".

Healthline, 2020. https://www.healthline.com/health/coronavirus-vs-sars#covid-19-vs-sars

Tesini, Brenda L. "Coronaviruses and Acute Respiratory Syndromes (COVID-19, MERS, and SARS)". Merck Manual, 2020. https://www.merckmanuals.com/en-ca/professional/infectious-diseases/respiratory-viruses/coronaviruses-and-acute-respiratory-syndromes-covid-19-mers-and-sars

Whiting, Kate. "Two experts explain what other viruses can teach us about COVID-19 - and what they can't". World Economic Forum, 2020. https://www.weforum.org/agenda/2020/03/coronavirus-covid-19-mers-sars-experts/

Works Cited (Chapter 3)

Abdul-Rasool, Sahar, and Burtram C. Fielding. "Understanding Human Coronavirus HCoV-NL63~!2009-11-13~!2010-04-09~!2010-05-25~!" The Open Virology Journal, vol. 4, no. 1, 25 May 2010, pp. 76–84., doi: 10.2174/1874357901004010076.

Auwaerter, Paul. "Coronavirus: Johns Hopkins ABX Guide." Coronavirus | Johns Hopkins ABX Guide, 10 June 2020, www.hopkinsguides.com/hopkins/view/Johns_Hopkins_ABX_Guide/540143/all/Coronavirus.

Cafasso, Jacquelyn. "Kawasaki Disease: Causes, Symptoms & Diagnosis." Edited by Kristi Pahr, Healthline, Healthline Media, 22 Oct. 2016, www.healthline.com/health/kawasaki-disease#takeaway.

CDC Contributers. "CDC SARS Response Timeline." Centers for Disease Control and Prevention, Centers for Disease Control and Prevention, 26 Apr. 2013, www.cdc.gov/about/history/sars/timeline.htm.

Chiu, S. S., et al. "Human Coronavirus NL63 Infection and Other Coronavirus

Infections in Children Hospitalized with Acute Respiratory Disease in Hong Kong, China." Clinical Infectious Diseases, vol. 40, no. 12, 2005, pp. 1721–1729., doi:10.1086/430301.

Cuffari, Benedette. "What Is OC43?" News Medical Life Sciences, 22 June 2020, www.news-medical.net/health/What-is-OC43.aspx.

Lau, S. K. P., et al. "Coronavirus HKU1 and Other Coronavirus Infections in Hong Kong." Journal of Clinical Microbiology, vol. 44, no. 6, 1 June 2006, pp. 2063–2071., doi:10.1128/jcm.02614-05.

Lauer, Stephen A., et al. "The Incubation Period of Coronavirus Disease 2019 (COVID-19) From Publicly Reported Confirmed Cases: Estimation and Application." Annals of Internal Medicine, vol. 172, no. 9, 5 May 2020, pp. 577–582., doi:10.7326/m20-0504.

Liu, Ding X., et al. "Human Coronavirus-229E, -OC43, -NL63, and -HKU1." Reference Module in Life Sciences, 7 May 2020, doi:10.1016/b978-0-12-809633-8.21501-x.

Mao, Ling, et al. "Neurologic Manifestations of Hospitalized Patients With Coronavirus Disease 2019 in Wuhan, China." JAMA Neurology, vol. 77, no. 6, 2020, p. 683., doi:10.1001/jamaneurol.2020.1127.

Marshall, William F. "Unusual Symptoms of Coronavirus: What Are They?" Mayo Clinic, Mayo Foundation for Medical Education and Research, 30 June 2020, www.mayoclinic.org/diseases-conditions/coronavirus/expert-answers/coronavirus-unusual-symptoms/faq-20487367.

Nicholls, John M, et al. "Time Course and Cellular Localization of SARS-CoV Nucleoprotein and RNA in Lungs from Fatal Cases of SARS." PLoS Medicine, vol. 3, no. 2, 3 Feb. 2006, doi:10.1371/journal.pmed.0030027.

Paules, Catharine I., et al. "Coronavirus Infections—More Than Just the Common Cold." Jama, vol. 323, no. 8, 23 Jan. 2020, p. 707., doi:10.1001/jama.2020.0757.

Peiris, Joseph S.m., et al. "The Severe Acute Respiratory Syndrome." New England Journal of Medicine, vol. 349, no. 25, 18 Dec. 2003, pp. 2431–2441., doi:10.1056/nejmra032498.

Petrosillo, N., et al. "COVID-19, SARS and MERS: Are They Closely Related?" Clinical Microbiology and Infection, vol. 26, no. 6, 26 June 2020, pp. 729–734., doi:10.1016/j.cmi.2020.03.026.

Poutanen, Susan M. "Human Coronaviruses." Principles and Practice of Pediatric Infectious Diseases, 2018, doi:10.1016/b978-0-323-40181-4.00222-x.

Ragab, Dina, et al. "The COVID-19 Cytokine Storm; What We Know So Far." Frontiers in Immunology, vol. 11, 16 June 2020, doi:10.3389/fimmu.2020.01446.

Ratini, Melinda. "Acute Respiratory Distress Syndrome (ARDS): Definition, Symptoms, and Treatment." WebMD, WebMD, 4 Jan. 2020, www.webmd.com/lung/ards-acute-respiratory-distress-syndrome.

Selner, Marissa. "Croup: Causes, Symptoms, and Diagnosis." Edited by Melanie Santos, Healthline, Healthline Media, 7 Jan. 2016, www.healthline.com/health/croup#TOC_TITLE_HDR_1.

Wadman, Meredith, et al. "How Does Coronavirus Kill? Clinicians Trace a Ferocious Rampage through the Body, from Brain to Toes." Science, AAAS, 28 May 2020, www.sciencemag.org/news/2020/04/how-does-coronavirus-kill-clinicians-trace-ferocious-rampage-through-body-brain-toes.

WHO Africa. "Coronavirus." World Health Organization, World Health Organization, www.afro.who.int/publications/

coronavirus#:~:text=Coronaviruses%20are%20zoonotic%2C%20
meaning,not%20yet%20infected%20humans.

Xu, Xiao-Wei, et al. "Clinical Findings in a Group of Patients Infected with
the 2019 Novel Coronavirus (SARS-Cov-2) Outside of Wuhan, China:
Retrospective Case Series." Bmj, 19 Feb. 2020, p. m606., doi:10.1136/
bmj.m606.

Works Cited (Chapter 4)

Buonavoglia, Canio, et al. "Canine Coronavirus Highly Pathogenic for Dogs."
Emerging Infectious Diseases, vol. 12, no. 3, 23 Feb. 2006, pp. 492–
494., doi:10.3201/eid1203.050839.

Choi, Charles Q. "Monkey DNA Points to Common Human Ancestor."
LiveScience, Purch, 12 Apr. 2007, www.livescience.com/1411-monkey-
dna-points-common-human-ancestor.html.

"COVID-19 and Animals." Centers for Disease Control and Prevention,
Centers for Disease Control and Prevention, 22 June 2020, www.cdc.
gov/coronavirus/2019-ncov/daily-life-coping/animals.html.

Daly, Natasha. "Seven More Big Cats Test Positive for Coronavirus at Bronx
Zoo." National Geographic, National Geographic, 22 Apr. 2020,
www.nationalgeographic.com/animals/2020/04/tiger-coronavirus-
covid19-positive-test-bronx-zoo/.

Deng, Wei, et al. "Primary Exposure to SARS-CoV-2 Protects against
Reinfection in Rhesus Macaques." Science, 2 July 2020, doi:10.1126/
science.abc5343.

Enserink, Martin. "Coronavirus Rips through Dutch Mink Farms, Triggering
Culls to Prevent Human Infections." Science, AAAS, 23 June 2020,
www.sciencemag.org/news/2020/06/coronavirus-rips-through-
dutch-mink-farms-triggering-culls-prevent-human-infections.

Fox, Maggie. "Pangolins May Have Incubated the Novel Coronavirus, Gene Study Shows." Coronavirus, CTV News, 31 May 2020, www.ctvnews. ca/health/coronavirus/pangolins-may-have-incubated-the-novel-coronavirus-gene-study-shows-1.4961699.

Gallagher, Alex. "Feline Enteric Coronavirus - Digestive System." Merck Veterinary Manual, Merck Veterinary Manual, June 2020, www. merckvetmanual.com/digestive-system/diseases-of-the-stomach-and-intestines-in-small-animals/feline-enteric-coronavirus.

Gollakner, Rania. "Canine Coronavirus Disease." vca_corporate, vcahospitals. com/know-your-pet/coronavirus-disease-in-dogs.

Gould, Kevin. "Calf Scours Signs, Treatment and Prevention: Part 2." MSU Extension, Michigan State University , 2 Oct. 2018, www.canr.msu. edu/news/calf_scours_signs_treatment_and_prevention_part_2.
Gruenberg, Walter. "Intestinal Diseases in Cattle - Digestive System." Merck Veterinary Manual, Merck Veterinary Manual, Aug. 2014, www. merckvetmanual.com/digestive-system/intestinal-diseases-in-ruminants/intestinal-diseases-in-cattle#v3263210.

Hu, Hui, et al. "Isolation and Characterization of Porcine Deltacoronavirus from Pigs with Diarrhea in the United States." Journal of Clinical Microbiology, vol. 53, no. 5, 20 Feb. 2015, pp. 1537–1548., doi:10.1128/ jcm.00031-15.

Hunter, Tammy. "Feline Infectious Peritonitis." VCA Corporate, vcahospitals. com/know-your-pet/feline-infectious-peritonitis.

Janzen, Eugene, and Murray Jelinski. "Bovine Respiratory Disease." Beef Cattle Research Council, 2 Oct. 2019, www.beefresearch.ca/research-topic.cfm/bovine-respiratory-disease-38.

Koonpaew, Surapong, et al. "PEDV and PDCoV Pathogenesis: The Interplay

Between Host Innate Immune Responses and Porcine Enteric Coronaviruses." Frontiers in Veterinary Science, vol. 6, 22 Feb. 2019, doi:10.3389/fvets.2019.00034.

Lee, Changhee. "Porcine Epidemic Diarrhea Virus: An Emerging and Re-Emerging Epizootic Swine Virus." Virology Journal, vol. 12, no. 1, 22 Dec. 2015, doi:10.1186/s12985-015-0421-2.

Marker, Teresa. "Managing Winter Dysentery In The Dairy Herd." Crystal Creek, 20 June 2016, crystalcreeknatural.com/managing-winter-dysentery-in-the-dairy-herd/.

Potter, Tim. "A Systematic Approach to Calf Gastroenteric Disease." Livestock, vol. 16, no. 2, 24 Mar. 2011, pp. 23–28., doi:10.1111/j.2044-3870.2010.00022.x.

"Protecting Great Apes from Covid-19." The Economist, The Economist Newspaper, 16 May 2020, www.economist.com/science-and-technology/2020/05/16/protecting-great-apes-from-covid-19.

Public Health Agency of Canada. "Government of Canada." Canada.ca, Canadian Government, 22 July 2020, www.canada.ca/en/public-health/services/diseases/2019-novel-coronavirus-infection/prevention-risks/animals-covid-19.html.

Saif, Linda J. "Bovine Respiratory Coronavirus." Veterinary Clinics of North America: Food Animal Practice, vol. 26, no. 2, 1 July 2010, pp. 349–364., doi:10.1016/j.cvfa.2010.04.005.

Shi, Jianzhong, et al. "Susceptibility of Ferrets, Cats, Dogs, and Other Domesticated Animals to SARS–Coronavirus 2." Science, vol. 368, no. 6494, 29 May 2020, pp. 1016–1020., doi:10.1126/science.abb7015.

Vlasova, A. N., et al. "Porcine Coronaviruses." Emerging and Transboundary Animal Viruses Livestock Diseases and Management, 23 Feb. 2020, pp. 79–110., doi:10.1007/978-981-15-0402-0_4.

Works Cited (Chapter 5)

Cui, Jie, et al. "Origin and Evolution of Pathogenic Coronaviruses." Nature
Reviews Microbiology, vol. 17, no. 3, 2018, pp. 181–192., doi:10.1038/
s41579-018-0118-9.

DiGiuseppe, Maurice, and Christine Adam-Carr. Nelson Biology 12: University
Preparation. Nelson Thomson Learning, 2012.

"Evolution of Viruses (Article) | Viruses." Khan Academy, Khan Academy,
www.khanacademy.org/science/biology/biology-of-viruses/virus-
biology/a/evolution-of-viruses.

Kenny, Vincent, et al. "Heuristic Algorithms." Optimization, 25 May 2014,
optimization.mccormick.northwestern.edu/index.php/Heuristic_
algorithms.

Loewe, Laurence. "Negative Selection." Nature News, Nature Publishing
Group, 2008, www.nature.com/scitable/topicpage/negative-
selection-1136/.

"Natural Selection." Natural Selection, evolution.berkeley.edu/evolibrary/
article/evo_25.

Prjibelski, Andrey D., et al. "Sequence Analysis." Encyclopedia of
Bioinformatics and Computational Biology, 2019, pp. 292–322.,
doi:10.1016/b978-0-12-809633-8.20106-4.

Saeed, Usman, and Zainab Usman. "Biological Sequence Analysis."
Computational Biology [Internet]., edited by H Husi, Codon
Publications, 2019, www.ncbi.nlm.nih.gov/books/NBK550342/.

"Viral Evolution." Nature News, Nature Publishing Group, 2020, www.nature.com/subjects/viral-evolution.

"Virus Strains." Science Learning Hub, 9 Dec. 2014, www.sciencelearn.org.nz/resources/184-virus-strains.

Wertheim, J. O., et al. "A Case for the Ancient Origin of Coronaviruses." Journal of Virology, vol. 87, no. 12, 2013, pp. 7039–7045., doi:10.1128/jvi.03273-12.

Wertheim, Joel O., and Sergei L. Kosakovsky Pond. "Purifying Selection Can Obscure the Ancient Age of Viral Lineages." Molecular Biology and Evolution, vol. 28, no. 12, 2011, pp. 3355–3365., doi:10.1093/molbev/msr170.

Zheng, Wen-Xin, et al. "Coronavirus Phylogeny Based on a Geometric Approach." Molecular Phylogenetics and Evolution, vol. 36, no. 2, 2005, pp. 224–232., doi:10.1016/j.ympev.2005.03.030.

Zheng, Xiaoqi, et al. "A Poisson Model of Sequence Comparison and Its Application to Coronavirus Phylogeny." Mathematical Biosciences, vol. 217, no. 2, 2009, pp. 159–166., doi:10.1016/j.mbs.2008.11.006.

Works Cited (Chapter 6)

Bernstein, Aaron, and Renee N. Salas. "Coronavirus and Climate Change." C-CHANGE | Harvard T.H. Chan School of Public Health, 6 July 2020, www.hsph.harvard.edu/c-change/subtopics/coronavirus-and-climate-change/.

Bukhari, Qasim, and Yusuf Jameel. "Will Coronavirus Pandemic Diminish by Summer?" SSRN Electronic Journal, 26 Mar. 2020, doi:10.2139/ssrn.3556998.

Chan, K H, et al. "The Effects of Temperature and Relative Humidity on the Viability of the SARS Coronavirus." Advances in Virology, Hindawi Publishing Corporation, www.ncbi.nlm.nih.gov/pubmed/22312351.

Chan, K. H., et al. "The Effects of Temperature and Relative Humidity on the Viability of the SARS Coronavirus." Advances in Virology, vol. 2011, pp. 1–7., doi:10.1155/2011/734690.

Goudarzi, Sara. "How a Warming Climate Could Affect the Spread of Diseases Similar to COVID-19." Scientific American, Scientific American, 29 Apr. 2020, www.scientificamerican.com/article/how-a-warming-climate-could-affect-the-spread-of-diseases-similar-to-covid-19/.

Gupta, Sonal, et al. "Effect of Weather on COVID-19 Spread in the US: A Prediction Model for India in 2020." Science of The Total Environment, vol. 728, 21 Apr. 2020, p. 138860., doi:10.1016/j.scitotenv.2020.138860.

Kenney, Scott, et al. "Coronaviruses, Spillovers, and Spilled Milk." Infectious Diseases Institute, 21 Feb. 2020, idi.osu.edu/news-articles/coronaviruses-spillovers-spilled-milk.

Khillar, Sagar. "Difference Between Absolute and Relative Humidity." Difference Between Similar Terms and Objects, 10 June 2019, www.differencebetween.net/science/nature/difference-between-absolute-and-relative-humidity/.

Ma, Yueling, et al. "Effects of Temperature Variation and Humidity on the Death of COVID-19 in Wuhan, China." Science of The Total Environment, vol. 724, 1 July 2020, p. 138226., doi:10.1016/j.scitotenv.2020.138226.

Maron, Dina. "'Wet Markets' Likely Launched the Coronavirus. Here's What You Need to Know." 'Wet Markets' Launched the Coronavirus. Here's

What You Need to Know., 16 Apr. 2020, www.nationalgeographic.
com/animals/2020/04/coronavirus-linked-to-chinese-wet-
markets/#close.

Moriyama, Miyu, and Takeshi Ichinohe. "High Ambient Temperature
Dampens Adaptive Immune Responses to Influenza A Virus
Infection." Proceedings of the National Academy of Sciences, vol. 116,
no. 8, 4 Aug. 2019, pp. 3118–3125., doi:10.1073/pnas.1815029116.

Pecl, Gretta T., et al. "Biodiversity Redistribution under Climate Change:
Impacts on Ecosystems and Human Well-Being." Science, vol. 355,
no. 6332, 30 Mar. 2017, doi:10.1126/science.aai9214.

Powell, Alvin. "Air Conditioning May Be Factor in COVID-19 Spread in the
South." Harvard Gazette, Harvard Gazette, 30 June 2020, news.
harvard.edu/gazette/story/2020/06/air-conditioning-may-be-factor-
in-covid-19-spread-in-the-south/.

Rajewski, Genevieve. "What Is Behind COVID-19 Spillover Events?" News
Center at Cummings School of Veterinary Medicine at Tufts
University, 29 May 2020, news.vet.tufts.edu/2020/05/what-is-
behind-covid-19-spillover-events/.

Ries, Julia. "Here's How COVID-19 Compares to Past Outbreaks." Healthline,
Healthline Media, 12 Mar. 2020, www.healthline.com/health-news/
how-deadly-is-the-coronavirus-compared-to-past-outbreaks.

Ryding, Sara. "What Is a Spillover Event?" News, 24 June 2020, www.news-
medical.net/health/What-is-a-Spillover-Event.aspx.

Stanford University. "Forest Loss Could Make Diseases like COVID-19 More
Likely, According to Study." World Economic Forum, Futurity , 15 Apr.
2020, www.weforum.org/agenda/2020/04/forest-loss-diseases-
covid19-coronavirus-deforestation-health.

Sung, Ming-Hua, et al. "Phylogeographic Investigation of 2014 Porcine Epidemic Diarrhea Virus (PEDV) Transmission in Taiwan." Plos One, vol. 14, no. 3, 6 Mar. 2019, doi:10.1371/journal.pone.0213153.

Xu, Hao, et al. "Possible Environmental Effects on the Spread of COVID-19 in China." Science of The Total Environment, vol. 731, 7 May 2020, p. 139211., doi:10.1016/j.scitotenv.2020.139211.

Works Cited (Chapter 7)

Appenzeller-Herzog, Christian, and Hans-Peter Hauri. "The ER-Golgi Intermediate Compartment (ERGIC): in Search of Its Identity and Function." Journal of Cell Science, The Company of Biologists Ltd, 1 June 2006, jcs.biologists.org/content/119/11/2173.

Astuti, Indwiani, and Ysrafil. "Severe Acute Respiratory Syndrome Coronavirus 2 (SARS-CoV-2): An Overview of Viral Structure and Host Response." Diabetes & Metabolic Syndrome: Clinical Research & Reviews, vol. 14, no. 4, 2020, pp. 407–412., doi:10.1016/j.dsx.2020.04.020.

Gelderblom, Hans R. "Structure and Classification of Viruses." Medical Microbiology. 4th Edition, University of Texas Medical Branch at Galveston., 1996.

Ghebreyesus, Tedros Adhanom. "WHO Director-General's Opening Remarks at the Media Briefing on COVID-19 - 11 March 2020." World Health Organization, World Health Organization, 11 Mar. 2020, www.who.int/dg/speeches/detail/who-director-general-s-opening-remarks-at-the-media-briefing-on-covid-19---11-march-2020.

Goulding, John. "Virus Replication." British Society for Immunology, www.immunology.org/public-information/bitesized-immunology/pathogens-and-disease/virus-replication.

"Intro to Viruses (Article)." Khan Academy, Khan Academy, www.
    khanacademy.org/science/high-school-biology/hs-human-body-
    systems/hs-the-immune-system/a/intro-to-viruses.

Kaiser, Gary. "10.2: Size and Shapes of Viruses." Biology LibreTexts, Libretexts,
    14 July 2020, bio.libretexts.org/Bookshelves/Microbiology/Book:_
    Microbiology_(Kaiser)/Unit_4:_Eukaryotic_Microorganisms_and_
    Viruses/10:_Viruses/10.02:_Size_and_Shapes_of_Viruses.

Li, Fang. "Structure, Function, and Evolution of Coronavirus Spike Proteins."
    Annual Review of Virology, vol. 3, no. 1, 2016, pp. 237–261.,
    doi:10.1146/annurev-virology-110615-042301.

Mallapaty, Smriti. "Why Does the Coronavirus Spread so Easily between
    People?" Nature News, Nature Publishing Group, 6 Mar. 2020, www.
    nature.com/articles/d41586-020-00660-x.

Petersen, Eskild, et al. "Comparing SARS-CoV-2 with SARS-CoV and Influenza
    Pandemics." The Lancet Infectious Diseases, 2020, doi:10.1016/s1473-
    3099(20)30484-9.

Wilde, Adriaan H. De, et al. "Host Factors in Coronavirus Replication." Roles
    of Host Gene and Non-Coding RNA Expression in Virus Infection
    Current Topics in Microbiology and Immunology, 2017, pp. 1–42.,
    doi:10.1007/82_2017_25.

Wu, Katherine J. "There Are More Viruses than Stars in the Universe. Why
    Do Only Some Infect Us?" National Geographic, 15 Apr. 2020, www.
    nationalgeographic.com/science/2020/04/factors-allow-viruses-
    infect-humans-coronavirus/.

Zimmer, Katarina. "Why Ro Is Problematic for Predicting COVID-19
    Spread." The Scientist Magazine®, 13 July 2020, www.the-scientist.
    com/features/why-ro-is-problematic-for-predicting-covid-19-

spread-67690.

Works Cited (Chapter 8)

Bronca, Tristan. "COVID-19: A Canadian Timeline." Canadian Healthcare Network, 8 Apr. 2020, www.canadianhealthcarenetwork.ca/covid-19-a-canadian-timeline.

Corum, Jonathan, et al. "Coronavirus Vaccine Tracker." The New York Times, The New York Times, 10 June 2020, www.nytimes.com/interactive/2020/science/coronavirus-vaccine-tracker.html?auth=login-google.

Cyranoski, David. "What China's Coronavirus Response Can Teach the Rest of the World." Nature News, Nature Publishing Group, 17 Mar. 2020, www.nature.com/articles/d41586-020-00741-x.

Elflein, John. "Canada: COVID-19 Tests and Deaths." Statista, 10 Aug. 2020, www.statista.com/statistics/1107034/covid19-cases-deaths-tests-canada/.

Gilmore, Rachel. "It's Now Recommended That Canadians Wear Face Masks." CTVNews, CTV News, 20 May 2020, www.ctvnews.ca/politics/it-s-now-recommended-that-canadians-wear-face-masks-1.4946752?cache=yes%3FclipId.

Hegarty, Stephanie. "The Chinese Doctor Who Tried to Warn Others about Coronavirus." BBC News, BBC, 6 Feb. 2020, www.bbc.com/news/world-asia-china-51364382.

"Reopening Ontario." Ontario.ca, 27 Apr. 2020, www.ontario.ca/page/reopening-ontario.

Rosenthal, Alex. "When Is A Pandemic Over?" YouTube, uploaded by TED-Ed,

1 Jun. 2020, https://www.youtube.com/watch?v=Qioedf_nJDo

Saba, Rosa. "Less than 4% of Canadians Have the COVID Alert Tracing App
- despite Better Privacy Protection than Facebook." Thestar.com, 5
Aug. 2020, www.thestar.com/business/2020/08/05/just-13-million-
canadians-have-downloaded-covid-alert-will-that-be-enough-to-
be-effective.html.

Stone, Laura, et al. "Ontario and Quebec Order Non-Essential Businesses to
Close." The Globe and Mail, 24 Mar. 2020, www.theglobeandmail.
com/politics/article-ontario-quebec-order-non-essential-
businesses-to-close/.

Yong, Story by Ed. "How the Pandemic Defeated America." The Atlantic,
Atlantic Media Company, 6 Aug. 2020, www.theatlantic.com/
magazine/archive/2020/09/coronavirus-american-failure/614191/.

Zhong, Raymond, and Paul Mozur. "To Tame Coronavirus, Mao-Style Social
Control Blankets China." The New York Times, The New York Times,
15 Feb. 2020, www.nytimes.com/2020/02/15/business/china-
coronavirus-lockdown.html.

Zhong, Raymond, and Vivian Wang. "China Ends Wuhan Lockdown, but
Normal Life Is a Distant Dream." The New York Times, The New
York Times, 7 Apr. 2020, www.nytimes.com/2020/04/07/world/asia/
wuhan-coronavirus.html.

"China." Worldometer, 2020, www.worldometers.info/coronavirus/country/
china/.

Works Cited (Chapter 9)

CDC. "Coronavirus Disease 2019 (COVID-19)." Centers for Disease Control and

Prevention, 11 Feb. 2020, https://www.cdc.gov/coronavirus/2019-ncov/hcp/planning-scenarios.html.

"For Some Israelis, Covid-19 Has Meant Separation from Their Families." J., 5 Aug. 2020, https://www.jweekly.com/2020/08/05/does-he-remember-me-for-some-israelis-covid-19-has-meant-separation-from-their-families/.

OECD. "Tourism Policy Responses to the Coronavirus (COVID-19)."Organization of Economic Co-operation and Development, 2 June 2020, https://www.oecd.org/coronavirus/policy-responses/tourism-policy-responses-to-the-coronavirus-covid-19-6466aa20/.

Wu, Jin, et al. "How the Virus Got Out." The New York Times, 22 Mar. 2020. NYTimes.com, https://www.nytimes.com/interactive/2020/03/22/world/coronavirus-spread.html.

Works Cited (Chapter 10)

Abdel-Qadir Husam, et al. "Importance of Considering Competing Risks in Time-to-Event Analyses." Circulation: Cardiovascular Quality and Outcomes, vol. 11, no. 7, American Heart Association, July 2018, p. e004580. ahajournals.org (Atypon), doi:10.1161/CIRCOUTCOMES.118.004580.

CDC. "Coronavirus Disease 2019 (COVID-19)." Centers for Disease Control and Prevention, 11 Feb. 2020. www.cdc.gov, https://www.cdc.gov/coronavirus/2019-ncov/need-extra-precautions/people-with-medical-conditions.html.

GISP, Lauren Bennett, PhD, Este Geraghty, MD, MS, MPH. "Mapping and Modeling Combine to Provide COVID-19 Forecasts." Esri, 7 Apr. 2020. www.esri.com, https://www.esri.com/about/newsroom/blog/models-maps-explore-covid-19-surges-capacity/.

Lancet, The. "Syndemics: Health in Context." The Lancet, vol. 389, no. 10072, Elsevier, Mar. 2017, p. 881. www.thelancet.com, doi:10.1016/S0140-6736(17)30640-2.

Lesko, Catherine R., and Bryan Lau. "Bias Due to Confounders for the Exposure-Competing Risk Relationship." Epidemiology (Cambridge, Mass.), vol. 28, no. 1, Jan. 2017, pp. 20–27. PubMed Central, doi:10.1097/EDE.0000000000000565.

Papastefanopoulos, Vasilis, et al. "COVID-19: A Comparison of Time Series Methods to Forecast Percentage of Active Cases per Population." Applied Sciences, vol. 10, no. 11, 11, Multidisciplinary Digital Publishing Institute, Jan. 2020, p. 3880.www.mdpi.com, doi:10.3390/app10113880.

Wolkewitz, Martin, and Livia Puljak. "Methodological Challenges of Analysing COVID-19 Data during the Pandemic." BMC Medical Research Methodology, vol. 20, no. 1, Apr. 2020, p. 81. BioMed Central, doi:10.1186/s12874-020-00972-6.

Zhou, Chenghu, et al. "COVID-19: Challenges to GIS with Big Data." Geography and Sustainability, vol. 1, no. 1, Mar. 2020, pp. 77–87. ScienceDirect, doi:10.1016/j.geosus.2020.03.005.

Works Cited (Chapter 11)

"Myths & Misinformation - COVID-19". University of Toronto Libraries, 2020. https://guides.library.utoronto.ca/c.php?g=715025&p=5097957

Ali, S. Harris. "#COVID19: Social media both a blessing and a curse during coroanvirus pandemic". The Conversation, 2020. https://theconversation.com/covid19-social-media-both-a-blessing-and-a-curse-during-coronavirus-pandemic-133596

Boutilier, Alex. "COVID-19 'infodemic' reaching Canadians through social media and apps, survey suggests". The Star, 2020. https://www. thestar.com/politics/federal/2020/05/12/facebook-says-it-flagged-covid-19-misinformation-50-million-times-in-april.html

Butler, Colin. "Ontario nurse under investigation after anti-vax, COVID conspiracy social media posts". CBC News, 2020. https://www.cbc.ca/news/canada/london/ontario-nurse-covid-19-conspiracy-social-media-1.5659338

Mosleh, Omar. "'This is misinformation on steroids': The Canadian who took on Gwyneth Paltrow is debunking coronavirus myths". The Star, 2020. https://www.thestar.com/news/canada/2020/04/06/the-canadian-who-took-on-gwyneth-paltrow-is-debunking-coronavirus-cures.html

Pazzanese, Christine. "Battling the 'pandemic of misinformation'". The Harvard Gazette, 2020. https://news.harvard.edu/gazette/story/2020/05/social-media-used-to-spread-create-covid-19-falsehoods/

Works Cited (Chapter 12)

"Analysing the mental well-being of adolescents and young adults during COVID-19". The European Commission, 2020. https://cordis.europa.eu/article/id/421549-analysing-the-mental-well-being-of-adolescents-and-young-adults-during-covid-19

Brown, David. "How COVID-19 changed consumer digital behaviour". The Canadian Grocer, 2020. https://www.canadiangrocer.com/top-stories/how-covid-19-changed-consumer-digital-behaviour-96680

Bowden, Olivia & Patel, Arti. "'A lot of fear': Asian community a target of racism amid coronavirus threat". Global News, 2020. https://globalnews.ca/news/6467500/coronavirus-racism/

Gold, Jessica. "Covid-19 Might Lead To A 'Mental Health Pandemic'". Forbes, 2020. https://www.forbes.com/sites/jessicagold/2020/08/06/covid-19-might-lead-to-a-mental-health-pandemic/#4d6dda60706f

Koeze, Ella & Popper, Nathaniel. "The Virus Changed the Way We Internet". The New York Times, 2020. https://www.nytimes.com/interactive/2020/04/07/technology/coronavirus-internet-use.html

Ley, David J. "Xenophobia in Response to Pandemics is (Sadly) Normal". Psychology Today, 2020. https://www.psychologytoday.com/ca/blog/women-who-stray/202002/xenophobia-in-response-pandemics-is-sadly-normal

Ma, Alexandra. "South Korea's 'Parasite' made Oscars history, but it still couldn't escape racist criticism". Insider, 2020. https://www.insider.com/parasite-oscars-wins-racist-criticism-coronavirus-2020-2

Park, Ed. "Confronting Anti-Asian Discrimination During the Coronavirus Crisis". The New Yorker, 2020. https://www.newyorker.com/culture/culture-desk/confronting-anti-asian-discrimination-during-the-coronavirus-crisis

Yuen, Kum Fai. "The Psychological Causes of Panic Buying Following a Health Crisis". International Journal of Environmental Research and Public Health, 2020. https://www.mdpi.com/1660-4601/17/10/3513

Works Cited (Chapter 13)

Borzykowski, Bryan. "Coronavirus Is Shaking up Canada's Housing Market, but Don't Expect a Crash." Macleans.ca, 26 May 2020, www.macleans.

ca/economy/realestateeconomy/coronavirus-is-shaking-up-
canadas-housing-market-but-dont-expect-a-crash/.

Collie, Meghan. "Can I Go to Another Province? The Latest Coronavirus Travel
Restrictions, by Region." Global News, Global News, 25 June 2020,
globalnews.ca/news/7106284/coronavirus-provincial-travel/.

"Concerns over COVID-19 Hit to Housing Market 'Overblown': Scotiabank
Economics." Scotiabank, 14 May 2020, www.scotiabank.com/ca/en/
about/perspectives.articles.economy.housing-economy.html.

"COVID-19 Pandemic Dramatically Shifts Canadians' Spending Habits."
Payments Canada, 13 May 2020, www.payments.ca/about-us/news/
covid-19-pandemic-dramatically-shifts-canadians%E2%80%99-
spending-habits.

Evans, Pete. "How COVID-19 Has Changed Canada's Economy for the
Worse - but Also for the Better | CBC News." CBCnews, CBC/Radio
Canada, 23 June 2020, www.cbc.ca/news/business/covid-economy-
changes-1.5618734.

Hensley , Laura. "How the Coronavirus Pandemic Has Put Canadian Tourism
in 'Survival Mode'." Global News, Global News, 7 June 2020,
globalnews.ca/news/7020442/tourism-canada-coronavirus/.

Hirchberg, Shimona. "The Impact of COVID-19 on Reading." BookNet
Canada, BookNet Canada, 15 Apr. 2020, www.booknetcanada.ca/
blog/2020/4/15/the-impact-of-covid-19-on-reading.

"How Canada's Tourism Industry Is Trying to Salvage Summer 2020."
CBCnews, CBC/Radio Canada, 28 June 2020, www.cbc.ca/radio/
costofliving/pandemic-aftermath-what-s-happening-to-tourism-
and-covid-driven-car-habits-in-canada-1.5628326/how-canada-s-
tourism-industry-is-trying-to-salvage-summer-2020-1.5630486.

Langton, James. "'Covid-19 Has Changed Everything,' StatsCan Says in Housing Market Study." Advisor's Edge, 21 July 2020, www.advisor.ca/news/economic/covid-19-has-changed-everything-statscan-says-in-housing-market-study/.

MacGregor, Sandra. "Air Canada Posts Loses Of 1.75 Billion, Urges Easing Of Travel Restrictions." Forbes, Forbes Magazine, 2 Aug. 2020, www.forbes.com/sites/sandramacgregor/2020/08/02/air-canada-posts-loses-of-175-billion-urges-easing-of-travel-restrictions/#5af5161f49bc.

"Reopening Ontario." Ontario.ca, Government of Ontario, 31 July 2020, www.ontario.ca/page/reopening-ontario.

Reynolds, Christopher. "COVID-19 Hits Tourism Businesses, Costing Canadian Cities Millions in Revenue." Global News, Global News, 28 Apr. 2020, globalnews.ca/news/6879282/coronavirus-business-tourism-canada-revenue/.

Schembri , Lawrence L. "Our COVID-19 Response: Navigating Diverse Economic Impacts." Bank of Canada, Bank of Canada, May 2020, www.bankofcanada.ca/2020/06/our-covid-19-response-navigating-diverse-economic-impacts/.

Statistics Canada. "Gross Domestic Product, Income and Expenditure, First Quarter 2020." Statistics Canada, Government of Canada, 29 May 2020, www150.statcan.gc.ca/n1/daily-quotidien/200529/dq200529a-eng.htm.

Statistics Canada. "Labour Force Survey, May 2020." Statistics Canada , Government of Canada, 5 June 2020, www150.statcan.gc.ca/n1/daily-quotidien/200605/dq200605a-eng.htm.

Statistics Canada. "Travel between Canada and Other Countries, December 2018." Statistics Canada, Government of Canada, 21 Feb. 2019, www150.statcan.gc.ca/n1/daily-quotidien/190221/dq190221c-eng.

htm.

Works Cited (Chapter 14)

CMHA. "COVID-19 Effects on the Mental Health of Vulnerable Populations." CMHA National, https://cmha.ca/documents/covid-mental-health-effects-on-vulnerable-populations.

Edmondson, Amy C., and Zhike Lei. "Psychological Safety: The History, Renaissance, and Future of an Interpersonal Construct." Annual Review of Organizational Psychology and Organizational Behavior, vol. 1, no. 1, 2014, pp. 23–43. Annual Reviews, doi:10.1146/annurev-orgpsych-031413-091305.

Government of Canada, Statistics Canada. The Daily — Canadian Perspectives Survey Series 1: Impacts of COVID-19. 8 Apr. 2020, https://www150.statcan.gc.ca/n1/daily-quotidien/200408/dq200408c-eng.htm.

Government of Canada, Statistics Canada. The Impact of COVID-19 on the Canadian Labour Market. 9 Apr. 2020, https://www150.statcan.gc.ca/n1/pub/11-627-m/11-627-m2020028-eng.htm.

Marshall, T., Stea, J., & Tanguay, R. Addressing the Opioid Crisis During COVID-19. 23 May 2020, https://www.medpagetoday.com/publichealthpolicy/opioids/86655.

Ornell, Felipe, et al. "'Pandemic Fear' and COVID-19: Mental Health Burden and Strategies." Brazilian Journal of Psychiatry, vol. 42, no. 3, Associação Brasileira de Psiquiatria, June 2020, pp. 232–35. SciELO, doi:10.1590/1516-4446-2020-0008.

Public Health Agency of Canada. (2018). Key Health Inequalities in Canada: A National Portrait. 2018, https://www.canada.ca/content/dam/

phacaspc/documents/services/publications/science-researchdata/6.
PerceivedMentalHealth_EN_final.pdf

United Nations. Policy Brief: The Impacts of COVID-19 on Women. 9
Apr. 2020, https://www.unwomen.org/- /media/headquarters/
attachments/sections/library/publications/2020/policy-brief-the-
impactof-covid-19-on-women-en.pdf?la=en&vs=1406

Works Cited (Chapter 15)

"COVID-19-Related Infodemic and Its Impact on Public Health: A
Global Social Media Analysis". The American Journal of Tropical
Medicine and Hygiene, 2020. https://www.ajtmh.org/content/
journals/10.4269/ajtmh.20-0812

"COVID-19 significant impacts health services for noncommunicable
diseases". World Health Organization, 2020. https://www.who.
int/news-room/detail/01-06-2020-covid-19-significantly-impacts-
health-services-for-noncommunicable-diseases

"Negative Impacts of Community-Based Public Health Measures During
a Pandemic (e.g., COVID-19) on Children and Families". Public
Health Ontario, 2020. https://www.publichealthontario.ca/-/media/
documents/ncov/cong/2020/06/covid-19-negative-impacts-public-
health-pandemic-families.pdf?la=en

Bronco, Tristan. "COVID-19: A Canadian timeline". Canadian Healthcare
Network, 2020. https://www.canadianhealthcarenetwork.ca/covid-
19-a-canadian-timeline

Goodyer, Jason. "Coronavirus: Will COVID-19 become a seasonal virus?".
Science Focus, 2020. https://www.sciencefocus.com/news/
coronavirus-will-covid-19-become-a-seasonal-virus/

Jackson, Hannah. "COVID-19: Can you be infected with coronavirus more than once?". Global News, 2020. https://globalnews.ca/news/6623287/coronavirus-multiple-infections/

Kanzawa, Mia & Spindler, Hilary. "Will Coronavirus Disease 2019 Become Seasonal?" The Journal of Infectious Diseases, 2020. https://academic.oup.com/jid/article/222/5/719/5860444

Ranney, Megan L. "Critical Supply Shortages - The Need for Ventilators and Personal Protective Equipment during the Covid-19 Pandemic". The New England Journal of Medicine, 2020. https://www.nejm.org/doi/full/10.1056/NEJMp2006141

Scharping, Nathaniel. "Could We Be Living With COVID-19 Forever?". Discover Magazine, 2020. https://www.discovermagazine.com/health/could-we-be-living-with-covid-19-forever

Stulpin, Caitlyn. "Seasonality will 'eventually'" play a role in COVID-19 transmission". Healio, 2020. https://www.healio.com/news/infectious-disease/20200501/seasonality-will-eventually-play-a-role-in-covid19-transmission

Works Cited (Chapter 16)

Bloomberg, Jason Gale. "Here's What We Know About Kids and Covid-19." The Washington Post, WP Company, 10 Aug. 2020, www.washingtonpost.com/business/heres-what-we-know-about-kids-and-covid-19/2020/08/10/ed6f724c-dac2-11ea-b4f1-25b762cdbbf4_story.html.

Donaghue, Erin. "2,120 Hate Incidents against Asian Americans Reported during Coronavirus Pandemic." CBS News, CBS Interactive, 2 July 2020, www.cbsnews.com/news/anti-asian-american-hate-incidents-up-racism/.

"Unemployment and Mental Health." Home, 2009, www.iwh.on.ca/
summaries/issue-briefing/unemployment-and-mental-health.

www.ingramcontent.com/pod-product-compliance
Lightning Source LLC
Chambersburg PA
CBHW020706270326
41928CB00005B/299